Phytoremediation

A Strategy to Clean up Environment

– *Authors* –

Dr. H.C. Lakshman
Ratna V. Airsang

P.G. Department of Studies in Botany,
(Microbiology Laboratory)
Karnatak University, Dharwad-580 003
Karnataka-India

2016

Daya Publishing House®

A Division of

Astral International Pvt. Ltd.
New Delhi - 110 002

Cataloging in Publication Data--DK
Courtesy: D.K. Agencies (P) Ltd. <docinfo@dkagencies.com>

Lakshman, H. C., 1954- **author.**
Phytoremediation: A strategy to clean up environment/ authors, Dr. H.C. Lakshman, Ratna V. Airsang.
        pages cm
        Includes bibliographical references and index.

        ISBN 9789351309178 (International Edition)

        1. Phytoremediation. 2. Pollution.   I. Airsang, Ratna, author. II. Title.

TD192.75.L35 2015   DDC 628.5   23

Published by   :   **Daya Publishing House®**
                     A Division of
                     Astral International Pvt. Ltd.
                     – ISO 9001:2008 Certified Company –
                     4760-61/23, Ansari Road, Darya Ganj
                     New Delhi-110 002
                     Ph. 011-43549197, 23278134
                     E-mail: info@astralint.com
                     Website: www.astralint.com

Laser Typesetting: **SSMG Computer Graphics,** Delhi - 110 084

Printed at   :   **Thomson Press India Limited**

PREFACE

"Phytoremediation as a cleansing tool"- Many vegetative species treat the land areas polluted by a variety of hazardous substances. Therefore, scientists have been forced to become very creative in developing effective transgenic phytoremediators. Not only had the land, water pollutants can also be removed better from plants. Plants have the capacity to withstand relatively high concentration of organic chemicals without toxic effects and they can absorb and convert chemicals quickly to less toxic metabolites. In addition, they stimulate the degradation of organic chemicals in the rhizosphere by the release of root exudates and enzymes, and build up organic carbon in the soil. The use of specially selected and engineered metal-accumulating plants for environmental clean-up is an emerging biotechnology.

The principal application of phytoremediation is for lightly contaminated soils and waters where the material to be treated is at a shallow or medium depth and the area to be treated is large. This will make agronomic techniques economical and applicable for both planting and harvesting. In addition, the site owner must be prepared to accept a longer remediation period. Plants that are able to decontaminate soils do one or more of the following: Plant uptake of contaminant from soil particles or soil liquid into their roots; Bind the contaminant into their root tissue, physically or chemically and; Transport the contaminant from their roots into growing shoots and prevent or inhibit the contaminant from leaching out of the soil.

Moreover, the plants should not only accumulate, degrade or volatilize the contaminants, but should also grow quickly in a range of different conditions and tend themselves to easy harvesting. A quite number of plants are employed in the absorption of toxic metals from contaminated soil. Metal contamination can be remediated by approaches involving green plants and their associated micro flora, soil amendments and agronomic techniques. Plants absorb toxic metals, from the soil and atmosphere. The possible adverse effects of heavy metal pollution and their phytotoxic effects have been reported by several workers. In recent time, the bioremediation became modern thought and safe practice in elimination of heavy metals from the environment, as plants are known to accumulate them.

Scientists all over have been in search of some innovative, eco-friendly and low cost alternative technologies. One of them is the phytoremediation, which include the use of plants to clean and cure the environment; and plants have been known for their property to absorb, accumulate and detoxify the impurities present in the soil, water and air through various physical, chemical and biological processes. Phytoremediation, a fast-emerging new technology for removal of toxic Heavy Metals (HMs), is cost-effective, non-intrusive and aesthetically pleasing. It exploits the ability of selected plants to remediate pollutants from contaminated sites. Plants have inter-linked physiological and molecular mechanisms of tolerance of HMs. High tolerance to HM toxicity is based on a reduced metal uptake or increased internal sequestration, which is manifested by interaction between a genotype and its environment.

The growing interest in molecular genetics has increased our understanding of mechanisms of HM tolerance in plants and many transgenic plants have displayed increased HM tolerance. Improvement of plants by genetic engineering, that is, by modifying characteristics like metal uptake, transport and accumulation and plant's tolerance to metals, open up new possibilities of phytoremediation. Either naturally occurring or genetically engineered plants are used for cleaning contaminated environments. Phytoremediation can be used to remove not only metals (for example, Ag, Cd, Co, Cr, Cu, Hg, Mn, Mo, Ni, Pb, Zn) but also radio nuclides (for examples, ^{90}Sr, ^{137}Cs, ^{239}Pu, ^{234}U, ^{236}U) and certain organic compounds. Phytoremediation has its own advantages and limitations. Recent progress in determining the molecular basis for metal accumulation and tolerance by hyper accumulators has been significant, and provides us with a strong basis to outline some strategies for achieving these goals. Meanwhile a lot of work is needed to commercialize this technology in India.

Plant pathologists, Microbiologists, Biochemists, Ecologists and Biotechnologists of the country have been involved in the researches on AM fungal colonized higher plants and how to make these as phytoremediation tools to remove the contaminants of nature and have gathered valuable information on the exploitation of these biotools for the welfare of the (plant communities) humankind. However, no effort has been made to bring all such informations together. In view of this, we took a tiny step towards the compilation of valuable research work of many concern environmentalists. We have taken utmost care to include all the relevant information in the volume.

We would like to express our sincere gratitude to the Dr. M.N. Sreenivasa, Professor and Head, Department of Agricultural Microbiology, University of Agricultural Science, Dharwad - 580005 for rendering his valuable time in writing foreword and suggestions while making final draft of the book. We greatly indebted to our family members for their kind cooperation while completing this book and also extend our gratitude to the persons who have encouraged and supported directly and indirectly to complete this task. We also extend our sincere thanks to Astral Publishing Company, New Delhi for timely publishing this book with keen interest.

Dr. H.C. Lakshman

Ratna V. Airsang

UNIVERSITY OF AGRICULTURAL SCIENCES, DHARWAD

Dr. M.N. Sreenivasa

Professor and Head
Dept. of Agricultural Microbiology
University of Agricultural Sciences,
Krishi Nagar, Dharwad-580 005
Email: mnsreenivasa@gamil.com

FOREWORD

India being agriculture oriented country contributed different types of food crops due to suitability of environment and soil condition. However, due to Green revolution several high yielding varieties have been developed and adopted for cultivation in our country. These high yielding varieties demand higher quantity of nutrients. This lead to the use of chemical fertilizers to boost crop yields. At that time, it was essential to go for higher crop production to meet the demand for food. Hence, agricultural technologists were aiming only for higher yields and never thought of ill effects of these chemical fertilizers on soil micro biota, biochemical activities, pollution of groundwater, soil and air. By the time this was realized, there was deterioration of soil health due to accumulation of toxic elements because of indiscriminate use of fertilizers and pesticides which lead to decrease in production per unit area. Since then Government of India introduced Organic Farming policy and encouraged farmers to use more of organic manures to improve soil fertility and productivity in addition to reduce environmental pollution.

Environmentalists were equally worried about soil pollutants as they are likely to enter food grains and cause health hazards in human beings, animals, birds etc. Several scientists developed different technologies to remove these soil pollutants .Among them Phytoremediation is fast developing since a decade in which plants are used to remove heavy metals from the soil. In the present book, authors have very well explained mechanisms of biosorption of heavy metals in different plant species - phytoextraction, rhizofiltration, phytostabilization, phytovolatalization and phytodegradation. In addition to this, use of microbial inoculants –*Rhizobium, Azotobacter* and Arbuscular mycorrhizal fungi in bioremediation has been well documented by authors. They have also discussed pros and cons of Phytoremediation with an excellent bibliography. Certainly the information given in this text book will be useful to researchers, scholars and environmentalists. I take this opportunity to congratulate the authors, Prof H. C. Lakshman and Smt. Ratna V. Airsang, for their excellent and fruitful efforts in bringing out this book.

M.N. Sreenivasa

CONTENTS

1
Phytoremediation

1.1 Introduction

Phytoremediation is a type of bioremediation that takes advantage of the natural processes of plants growth in soil. The generic term 'Phytoremediation consists of the Greek prefix 'phyto' meaning plant, attached to the Latin term 'remedium' to correct or remove an evil (Cunningham *et al.*, 1996). This technology can be applied to both organic and inorganic pollutants present in soil (soil substrate), water (liquid substrate) or the air (Salt *et al.*, 1998). Phytoremediation is described as a natural process carried out by plants and trees in the cleaning up and stabilization of contaminated soils, sediments and ground water or it is described as the use of vegetation for *in situ* treatment of contaminated soils, sediments and water. Plants have shown the capacity to withstand relatively high concentration of organic chemicals without toxic effects and they can absorb and convert chemicals quickly to less toxic metabolites. In addition, they stimulate the degradation of organic chemicals in the rhizosphere by the release of root exudates and enzymes, and build up organic carbon in the soil. The use of specially selected and engineered metal-accumulating plants (Fig.1) for environmental clean-up is an emerging biotechnology, which include the following:

(i) Phytoextraction

Phytoextraction is the use of metal accumulating plants to remove toxic metals from soil. The plants absorb metal contaminants through the roots and move them into the stems and the leaves. The plants are harvested and then suitably discarded.

(ii) Rhizofiltration

Rhizofiltration is the use of plant roots to remove toxic metals from polluted waters and this process concentrates on surface and groundwater remediation whereby the roots directly absorb the contaminants or indirectly, by attracting and containing them. When the roots become saturated, the plants are harvested and then replaced. An example of this process is the use of constructed wetlands to treat wastewater and landfill leachate.

(iii) Phytostabilization

Phytostabilization is the use of plants to limit the bioavailability of toxic metals in soil. In this, plants are used to reduce the mobility of soil or water contaminants and contain them in a limited area. This can be accomplished by the plant roots, as detailed above or by decreasing soil erosion and wind-blown dust. For metals, which do not biodegrade, this is an effective method for preventing their spread to ground or surface water.

Phytoremediation of heavy metals is designed to concentrate metals in plant tissues from the solid or liquid heavy-metal-laden waste and developing an economical method of extracting the metals from plant residues. This is economical and eco-friendly as it will eliminate the need of costly off-site disposal of the sludge contaminating the ground water.

Types of Phytoremediation for organic metals:

i) Phytotransformation

Phytotransformation is through the plants own metabolic processes organic compounds in contaminated soils and groundwater are absorbed and degraded.

ii) Rhizosphere Bioremediation

Microorganisms are found in the area surrounding the roots of plants (rhizosphere) and are stimulated by natural substances released by the plant. These microorganisms aid the breakdown of the contaminants through their own metabolic processes.

Phytoremediation uses the natural ability of plants to contain, break down, or eliminate toxic chemicals and pollutants from soil or water. This process can be applied to clean up metals, pesticides, solvents, explosives, crude oil, and contaminants that are potentially harmful to the environment. Plants absorb and metabolize certain organic chemicals and metals, some of which are essential for plant growth, including zinc, copper, and iron. Some plants, absorb more metal than others, including metals that appear unnecessary for plant function, these are known as hyperaccumulators.

1.2 Sorption and Biosorption

Sorption

The cell wall of bacteria, yeast, algae and fungi which contains several active groups of constituents like acetamido group of chitin, polysaccharide amine (amino peptidoglycoside), sulphahydral and carboxyl groups in protein, phosphodiester (teichoic acid), phosphate and hydroxyl in polysaccharides, participate in biosorption (Gadd, 2000). The use of freely suspended microbial biomass suffers with disadvantages like small particle size, low mechanical strength and difficulty in separating biomass and effluent. The free cells can provide valuable information in laboratory experimentation but are not suited to industrial applications. Immobilized biomass offers many advantages including better reusability, high biomass loading and minimum clogging in continuous flow systems. Biomass immobilized in a range of inert materials like, silica, polyacrylamide, polymethane and polysulphone has been used in a variety of bioreactor configurations, including rotating biological contractors, fixed reactors, trickle filters, fluidized beds and air lift bioreactors (Gadd and White, 1993).

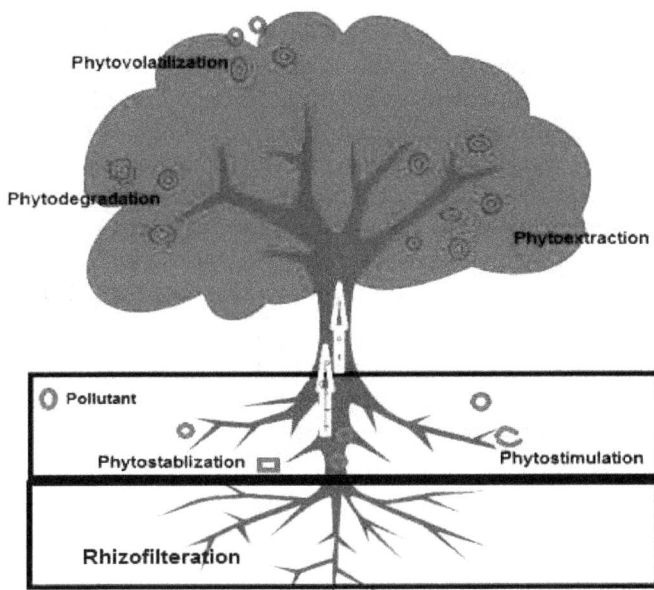

Fig. 1: Shows the process of uptake of pollutants by plant

Biosorption

Biosorption is a physiochemical process that occurs naturally in certain biomass which allows it to passively concentrate and bind contaminants onto its cellular structure (Volesky and Bohumil, 1990). Though using biomass in environmental cleanup has been in practice for a while, scientists and engineers are hoping this phenomenon will provide an economical alternative for removing toxic heavy metals from industrial wastewater and aid in environmental remediation.

Pollution interacts naturally with biological systems. It is currently uncontrolled, seeping into any biological entity within the range of exposure. The most problematic contaminants include heavy metals, pesticides and other organic compounds which can be toxic to wildlife and humans in small concentration. There are existing methods for remediation, but they are expensive or ineffective (Ahalya *et al.*, 2003). However, an extensive body of research has found that a wide variety of commonly discarded waste including eggshells, bones, peat, (Schildmeyer, Wolcott and Bender, 2009) fungi, seaweed, yeast and carrot peels (Bhatti *et al.*, 2010) can efficiently remove toxic heavy metal ion from contaminated water. Ions from metals like mercury can react in the environment to form harmful compounds like methylmercury a compound known to be toxic in humans. In addition, adsorbing biomass, or biosorbents, can also remove other harmful metals like: arsenic, lead, cadmium, cobalt, chromium and uranium (Lesmana *et al.*, 2009; Velásquez and Dussan, 2009). Biosorption may be used as an environmentally friendly filtering technique. There is no doubt that the world could benefit from more rigorous filtering of harmful pollutants created by industrial processes and all-around human activities.

The idea of using biomass as a tool in environmental cleanup has been around since the early 1900s when Arden and Lockett discovered certain types of living bacteria cultures were capable of recovering nitrogen and phosphorus from raw sewage when it was mixed in an aeration tank (Lesmana *et al.*, 2009; Alleman *et al.*, 1983). This discovery became known as the activated sludge process which is structured around the concept of bioaccumulation and is still widely used in wastewater treatment plants today. It wasn't until the late 1970s when scientists noticed the sequestering characteristic in dead biomass which resulted in a shift in research from bioaccumulation to biosorption (Lesmana *et al.*, 2009).

Biosorption is a metabolically passive process, meaning it does not require energy, and the amount of contaminants a sorbent can remove is dependent on kinetic equilibrium and the composition of the sorbents cellular surface. Contaminants are adsorbed onto the cellular structure.

Bioaccumulation is an active metabolic process driven by energy from a living organism and requires respiration. Bioaccumulation occurs by absorbing contaminants which are transferred onto and within the cellular surface.

Both bioaccumulation and biosorption occur naturally in all living organisms however, in a controlled experiment conducted on living and dead strains of *Bacillus sphaericus* it was found that the biosorption of chromium ions was 13–20% higher in dead cells than living cells. In terms of environmental remediation, biosorption is preferable to bioaccumulation because it occurs at a faster rate and can produce higher concentrations. Since metals are bound onto the cellular surface, biosorption is a reversible process whereas bioaccumulation is only partially reversible.

Biosorption of some Heavy Metals in Different Plant Species

Plants absorb toxic metals, both from the soil and atmosphere. They may enter the plants either by root system or through foliar uptake. Stunted growth, chlorosis, necrosis, leaf epinasty and red brownish discoloration are some of the visible symptoms of metal phytotoxicity. Plants represent an important pathway for the movement of potentially toxic trace elements from soil to human beings. The possible adverse effects of heavy metal pollution and their phytotoxic effects have been reported by several workers (Antonovies *et al.*, 1971, Chiba & Takahashi 1977, Heale & Ormrod 1983, Leblova *et al.*, 1986). All the heavy metals are potentially toxic at elevated concentrations (Gadd & White, 1989). Accumulation of heavy metals in plant parts showed inhibitory and promotory effect on growth. In recent time, the bioremediation became modern thought and safe practice in elimination of heavy metals from the environment, as plants are known to accumulate them. Water hyacinth (*Eichhornia crassipes*) is used for pollution treatment and is reported to remove heavy metals (Chigbo *et al.*, 1982; Selvapathy & Sreedhar 1991; Santosh and Dhandapani, 2013) like Ni, As, Cd, Pb, Hg, Cu, Mn, Cr and Zn. Keeping in view of the role of plants in elimination of heavy metals from soil, the present study aims to evaluate the amount of metals absorbed by the whole plant and accumulation in different plant parts (Fig. 2) to determine their efficacy or potentiality in bioremediation and also to assess their residues present in edible parts to see their hazardous level for consumption (Table 1 and 2).

Table 1: Accumulation of heavy metals at lower concentrations (in ppm).

Plant species	Hg		Cd		Zn		Cu	
	Stem	Root	Stem	Root	Stem	Root	Stem	Root
Solanum melongena L.	BDL	BDL	0.273	0.366	3.22	4.42	0.28	0.3
Zea mays L.	BDL	BDL	0.826	0.911	0.04	0.12	0.14	0.2
Tagetes erecta L.	BDL	BDL	0.693	0.712	22.7	24.2	3.7	4.2
Abelmoschus esculentum (L.) Moench	BDL	BDL	0.723	0.816	2.3	2.6	0.667	0.91

BDL: Below detectable limit

Table 2: Accumulation of heavy metals at lower concentrations (in ppm).

Plant species	Hg		Cd		Zn		Cu	
	Stem	Root	Stem	Root	Stem	Root	Stem	Root
Solanum melongena L.	BDL	BDL	0.350	0.412	6.12	7.42	0.42	0.5
Zea mays L.	BDL	BDL	0.912	0.966	0.11	0.16	0.39	0.4
Tagetes sps.	BDL	BDL	0.732	0.826	30.1	31.2	4.1	4.5
Abelmoschus esculentus (L.) Moench	BDL	BDL	3.21	4.12	29.47	30.4	0.775	1.22

BDL: Below detectable limit

Heavy Metals

Cadmium   Stem - *Zea mays > Abelmoschus esculentus > Tagetes > Solanum melongena*

Zinc   Root - *Zea mays >Abelmoschus esculentus > Tagetes > Solanum melongena*
Stem - *Tagetes > Solanum melongena > Abelmoschus esculentus > Zea mays*

Copper   *Tagetes > Solanum melongena > Abelmoschus esculentus > Zea mays*
Stem - *Tagetes > Abelmoschus esculentus > Solanum melongena > Zea mays*
Root - *Tagetes > Abelmoschus esculentus > Solanum melongena > Zea mays*

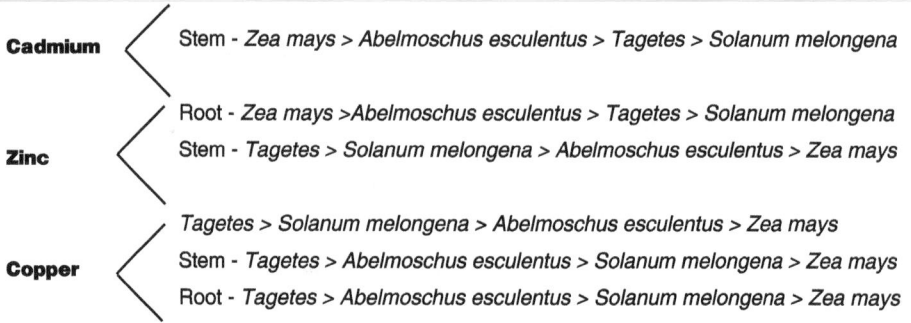

Fig.2: **Schematic representation of order of accumulation of heavy metals in some plant species when treated at lower concentrations.**

2

Mechanism of Metals Uptake by Plants

Bioavailability of metals is the primary factor responsible for the uptake of metals. In soils, metals exist as a variety of chemical forms in a dynamic equilibrium governed by the physical, chemical and biological processes of the soil. Bioavailability of soil pollutants, a primary basis of remediation efficacy, refers to a fraction of the total pollutant mass in the soil and sediment available to plants. Uptake of metals by plants involves root interception of metal ions, entry of metal ions into roots and their translocation to the shoot through mass flow and diffusion.

Plants have evolved highly specific mechanisms to take up, translocate, and store these nutrients. For example, metal movement across biological membranes is mediated by proteins with transport functions. In addition, sensitive mechanisms maintain intracellular concentration of metal ions within the physiological range. In general, the uptake mechanism is selective and plants preferentially acquired some ions over others. Ion uptake selectivity depends upon the structure and properties of membrane transporters. These characteristics allow transporters to recognize, bind and mediate the Trans membrane transport of specific ions. For example, some transporters mediate the transport of divalent cations, but do not recognize mono or trivalent ions. Hyperaccumulator plants do not only accumulate high levels of essential micronutrients, but can also absorb significant amounts of non-essential metals such as Cd.

The apoplast continuum of root epidermis and cortex is readily permeable for solutes. Apoplastic pathway is relatively unregulated, because water and dissolved substance can flow and diffuse without crossing the membrane. The cell walls of the endodermal layer act as a barrier for apoplastic diffusion into the vascular system. Apoplastic transport is limited by high cation exchange capacity (CEC) of the cell wall. In the symplastic transport, metal ions move across the plasma membrane, which usually has a large negative resting potential of approximately 170 mV (negative inside the membrane). This membrane potential provides a strong electrochemical gradient for the inward movement of the metal ions. Most metal ions enter plant cells by an energy-dependent process through specific or generic metal-ion carriers or channels. On entry into the roots, metal ions can either be stored in the root or forwarded to the shoot, primarily, through the xylem. The rate of metal translocation to the shoot may depend on metal concentration in the root. A phytochelatin (PC) - mediated metal binding in the xylem sap as a possible mechanism for metal translocation has been proposed. Nutrients destined for the developing cereal grain encounter several restricting barriers on their path towards their final storage sites in the grain. In order to identify transporters and chelating agents that may be involved in transport and deposition of Zn in the barley grain, expression profiles have been generated of four different tissue types; the transfer cells, the aleurone layer, the endosperm, and the embryo (Tauris *et al.*, 2009).

The Process of Phytoremediation

It includes the steps as follows;

2.1 Phytoextraction

Phytoextraction or phytomining is the process of planting a crop of a species that is known to accumulate contaminants in the shoots and leaves of the plants, and then harvesting the crop and removing the contaminant from the site. Unlike the destructive degradation mechanisms, this technique yields a mass of plant and contaminant (typically metals) that must be transported for disposal or recycling. This is a concentration technology that leaves a much smaller mass to be disposed of when compared to excavation and landfilling. This technology is being evaluated in a Superfund Innovative Technology Evaluation (SITE) demonstration, and may also be a technology amenable to contaminant recovery and recycling. Phytoextraction is the name given to the process where plant roots absorb metal contaminants from the soil and translocate them to their above soil tissues. Phytoextraction also called phytoaccumulation refers to the uptake of metals from soil by plant roots into above ground portions of plants (Figure 3).

The concept of using plants to clean up contaminated environments is very old and cannot be traced to any particular source (Blaylock and Huang, 2000). Chaney (1983) was the first to reintroduce it as a remediation technique on metal-contaminated soils. Initially, the concept was based on metal hyper accumulating plants, which are able to uptake and tolerate extremely high levels of metals. In the past, extensive research has been conducted in the field of phytoextraction, searching for new phytoextractors (Baker and Brooks, 1989); providing more fundamental knowledge about metal uptake, translocation, and tolerance by plants (Rauser, 1995; Kramer *et al.*, 1996; Lasat *et al.*, 1998; Salt *et al.*, 1999) as well as improving plant metal accumulation and tolerance by genetic transformations (Karenlampi *et al.*, 2000;

Clemens *et al.*, 2002; Kramer, 2005). Another approach in the concept's development was based on high biomass-producing plants used together with chemical agents to enhance metal solubility and uptake by plants (Blaylock *et al.*, 1997; Huang *et al.*, 1997). Certain plants, called hyper-accumulators, absorb unusually large amounts of metals in comparison to other plants. More than 400 plant species have been identified to have potential for soil and water remediation (Lone *et al.*, 2008).

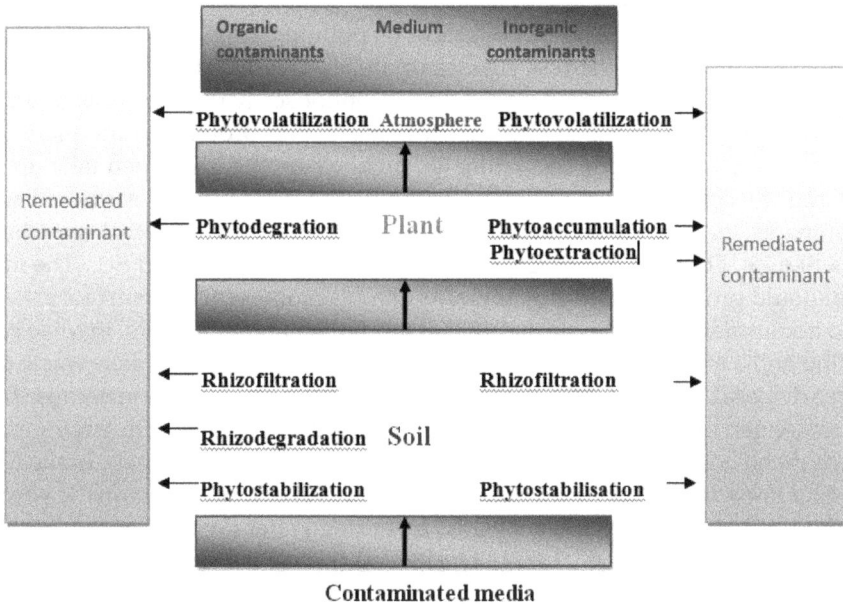

Fig.3: Flow diagram showing the process of Phytoremediation

Metal phytoextraction involves:

1. Cultivation of the appropriate plant/crop species on the contaminated site

2. Removal of harvestable metal-enriched biomass from the site and

3. Post-harvest treatments (that is, composting, compacting, thermal treatments) to reduce the volume and/or weight of biomass for disposal as a hazardous waste or for its recycling to reclaim valuable metals.

Two basic strategies of metal phytoextraction have been suggested, continuous or natural phytoextraction and induced, enhanced or chemically assisted phytoextraction (Salt *et al.*, 1998). After the plants have been allowed to grow for some time, they were harvested and either incinerated or composted to recycle the metals. This procedure may be repeated as necessary to bring soil contaminant levels down to allowable limits. If plants are incinerated, the ash must be disposed of in a hazardous waste landfill, but the volume of ash will be less than 10% of the volume that would be created if the contaminated soil itself were dug up for treatment. In some cases, it is possible to recycle the metals through a process known as phytomining, though this is usually reserved for use with precious metals.

2.2 Rhizofiltration

Rhizofiltration ('rhizo' means 'root') is the adsorption or precipitation on to plant roots (or absorption into the roots) of contaminants that are in solution surrounding the root zone. It is defined as the use of plants, both terrestrial and aquatic, to absorb, concentrate, and precipitate contaminants from polluted aqueous sources with low contaminant concentration in their roots. Rhizofiltration is similar to Phytoextraction but is concerned with the remediation of contaminated groundwater rather than the remediation of polluted soils. The contaminants are either adsorbed onto the root surface or are absorbed by the plant roots. The plants to be used for clean-up are raised in greenhouses with their roots in water. Contaminated water is collected from a waste site and brought to the plants, or the plants are planted in the contaminated area, where the roots then take up the water and the contaminants dissolved in it. As the roots become saturated with contaminants, they are harvested and disposed of safely. Rhizofiltration remediates metals like As, Pb, Cd, Ni, Cu, Cr, V and radionuclides (U, Cs and St). The ideal plants should produce significant amounts of roots biomass or root surface area, be able to accumulate and tolerate significant amounts of target metals, involve easy handling and a low maintenance cost and has a minimum of secondary waste that requires disposal. Terrestrial plants are more suitable for rhizofiltration because they produce longer, more substantial and often fibrous root systems with large surface areas or metal adsorption. *Pteris vittata,* commonly known as Chinese brake fern, is the first known As-hyper accumulator (Ma *et al.*, 2001). Several aquatic species have the ability remove HMs from water, including Water Pennywort (*Hydrocotyle umbrellate* L.) (Dierberg *et al.*, 1987). Duckweed (*Lemna minor* L.) (Mo *et al.*, 1989) and water hyacinth (*Eichhornia crassipes* (Mart.) Solms) (Zhu *et al.*, 1999). Indian mustard (*Brassica juncea*) and sunflower (*Helianthus annuus*) are most promising for metal removal from water. Indian mustard effectively removes Cd, Cr, Cu, Ni, Pb and Zn (Dushenkov *et al.*, 1995) whereas sunflower absorbs Pb (Dushenkov *et al.*, 1995) and U (Dushenkov *et al.*, 1997) from hydroponic solutions. Indian mustard could effectively remove a wide range (4 to 500 mg/L) of Pb concentration (Raskin and Ensley, 2000). Karkhanis *et al.*, (2005) reported the result of their experiment conducted on rhizofiltration under greenhouse condition using pistia, duckweed and water hyacinth (*E. crassipes)* to remediate aquatic environment contaminated by coal ash containing HMs. The results showed that pistia has high potential capacity of uptake of the HMs (Zn, Cr, and Cu) and duckweed also showed good potential for uptake of these metals next to pistia. Rhizofiltration of Zn and Cu in case of water hyacinth was lower as compared to pistia and duckweed. In a recent study, the potential of water hyacinth (*E. crassipes)* weeds for phytoremediation of metal polluted soils by rhizofiltration method was reported by Mohanty and Patra (2011).

2.3 Phytostabilization

Phytostabilization, also referred to as in-place inactivation, is primarily used for the remediation of soil, sediment, and sludges (United States Protection Agency, 2000). It is the use of plant roots to limit contaminant mobility and bioavailability in the soil and water. Contaminants are absorbed and accumulated by roots, adsorbed onto the roots, or precipitated in the rhizosphere. This reduces or even prevents the mobility of the contaminants preventing migration into the groundwater or air, and also reduces the bioavailability of the contaminant thus preventing spread through the food chain. This technique can also be used to re-establish a plant community on sites that have been denuded due to the high levels of metal contamination. Once a community of tolerant species has been established, the potential for wind erosion (and thus spread of the pollutant) is reduced and leaching of the soil contaminants is also reduced. The plants primary purposes are to:

(1) Decrease the amount of water percolating through the soil matrix, which may result in the formation of a hazardous leachate;

(2) Act as a barrier to prevent direct contact with the contaminated soil; and

(3) Prevent soil erosion and the distribution of the toxic metal to other areas (Raskin and Ensley, 2000).

Phytostabilization can occur through the sorption, precipitation, complexation or metal valence reduction. It is useful for the treatment of Pb as well as As, Cd, Cr, Cu and Zn. Some of the advantages associated with this technology are that the disposal of hazardous material/biomass is not required (United States Protection Agency, 2000) and it is very effective when rapid immobilization is needed to preserve ground and surface waters. The presence of plants also reduces soil erosion and decreases the amount of water available in the system (United States Protection Agency, 2000). Phytostabilization has been used to treat contaminated land areas affected by mining activities and superfund sites.

Smith and Bradshaw (1992), have developed two cultivars of *Agrostis tenius* and one of *Festuca rubra*, which are used for phytoremediation of the Pb, Zn and Cu contaminated soils. Phytostabilization, though most effective at sites having fine-textured soils with high organic matter content, can treat a wide range of surface contamination (Cunningham *et al.*, 1995; Berti and Cunningham, 2000).

Deep rooting plants could reduce the highly toxic Cr VI to Cr III, which are much less soluble and therefore, less bioavailable (James, 2001). Phytostabilization does not require soil removal and/or disposal of the hazardous material or the biomass. An experiment was conducted under green house condition using sorghum (fibrous root grass) to remediate soil contaminated by HMs and the developed vermicompost was amended in contaminated soil as a natural fertilizer (Jadia and Fulekar, 2008).

2.4 Phytovolatilization

The contaminant may become modified along the way, as the water travels along the plant's vascular system from the roots to the leaves, whereby the contaminants evaporate or volatize into the air surrounding the plant. The use of phytoextraction and phytovolatilization of metals by plants offers a viable remediation on commercial projects (Sakakibara *et al.*, 2007). Phytovolatization has been primarily used for the removal of mercury; the mercuric ion is transformed into less toxic elemental Hg (Ghosh and Singh, 2005). Phytovolatization has been successful in tritium (3H), a radioactive isotope of hydrogen; it is decayed to stable Helium with a half-life of about 12 years. Phytovolatilization is the most controversial of all phytoremediation technologies. Some metals, like As, Hg and Se, may exist as gaseous state in the environment. Some naturally occurring or genetically modified plants, like *Chara canescens* (musk grass), *Brassica juncea* (Indian mustard) and *Arabidopsis thaliana*, are reported to possess capability to absorb HMs and convert them to gaseous state within the plant and subsequently release them into the atmosphere (Ghose and Singh, 2005).

Unlike other remediation techniques, once the contaminants have been removed via volatilization, one has no control over their migration to other areas. Similar cases of volatization based soil remediation have also been reported in many recently published reports (Tangahu *et al.*, 2011; Conesa *et al.*, 2012).

2.5 Phytodegradation / phytotransformation

Phytodegradation is the breakdown of organic contaminants within plant tissue. Plants produce enzymes, such as dehydrogenase and oxygenase that help catalyze degradation. It appears that both the plants and the associated microbial communities play a significant role in attenuating contaminants. It is referred to the degradation or breakdown of organic contaminants by internal and external metabolic processes driven by the plant (Prasad and Freitas, 2003). The metabolic processes of microbes hydrolyse organic compounds into smaller units that can be absorbed by the plant. Some contaminants can be absorbed by the plant and are then broken down by plant enzymes. These smaller pollutant molecules may then be used as metabolites by the plant as it grows, thus becoming incorporated into the plant tissues and AM fungal association play role (Lakshman, 1994). Plant enzymes have been identified that breakdown ammonium wastes, chlorinated solvents such as TCE (Trichloroethylene), and others which degrade organic herbicides. Plant enzymes that metabolise contaminants may be released into the rhizosphere, where they may play active role in transformation of contaminants. Enzymes, like dehalogenase, nitro-reductase, peroxidase, laccase and nitrilase have been discovered in plant sediments and soils. Organic compounds such as munitions, chlorinates solvents, herbicides and insecticides and the inorganic nutrients can be degraded by this technology (Schnoor *et al.*, 1995). The dissolved TNT (trinitrotoluene) concentrations in flooded soil decreased from 128 ppm within one week in the presence of the aquatic plant, *Myriophyllum aquaticum*, which produces nitroreductase enzyme that can partially degrade TNT (Schnoor *et al.*, 1995). To engineer plant tolerance to TNT, two bacterial enzymes (PETN reductase and nitroreductase), able to reduce TNT into less harmful compounds, were over-expressed into less harmful compounds, were over-expressed in tobacco plants. The two genes *onr* and *nfs*, under the control of a constitutive promoter, provided the transgenic plants with increased tolerance to TNT at a concentration that severely affected the development of wild type plants (Hannink *et al.*, 2001).

Research related to this relatively new technology needs to be promoted and emphasized and expanded in developing countries since it is low cost. In situ, solar driven technology makes use of vascular plants to accumulate and translocate metals from roots to shoots. Harvesting the plant shoots can permanently remove these contaminants from the soil. Phytoremediation does not have the destructive impact on soil fertility and structure that some more vigorous conventional technologies have such as acid extraction and soil washing. This technology can be applied "in situ" to remediate shallow soil, ground water and surface water bodies. Also, phytoremediation has been perceived to be a more environmentally-friendly "green" and lowtech alternative to more active and intrusive remedial methods. The broader importance of protecting soils and improved management for the services they provide are currently receiving considerable attention from policy-makers. Soils provide fundamental ecosystem services, with extensive economic, ecological and sociological influences on the well being of the human society. Metal contaminated soils provide a significant but previously neglected component of the global soil resource. There is much scope to optimize the utilization of this resource for improved services. Phytoremediation does have real applications, but it is vital that it emerges as a realistic technology and in the right context. It has been tested successfully in many places around the world for many different contaminants (Table 3 and 4).

Table 3: Extent of testing phytoremediation across some sites in USA.

Location	Application	Pollutant	Medium	Plants
Ogden, UT	Phytoextraction and rhizodegradation	Petroleum and hydrocarbons	Soil and groundwater	Alfalfa, poplar, juniper, fescue
Anderson, ST	Phytostabilisation	HMs	Soil	Hybrid poplar, grasses
Ashtabula, OH	Rhizofiltration	Radionuclides	Groundwater	Sunflowers
Upton, NY	Phytoextraction	Radionuclides	Soil	Indian mustard, cabbage
Milan, TN	Phytodegradation	Expolsives waste	Groundwater	Duckweed, parrot feather
Amana, IA	Riparian corridor, Phytodegradation	Nitrates	Groundwater	Hybrid poplar
Pennysylvania	Phytoextraction mine wastes	Zinc and cadmium	Soil	Thlaspi caerulescens
SanFrancisco,CA	Phytovolatization	Se	Refinery wastes and agricultural soils	Brassica sp.

Table 4. Details of application of Phytoremediation

Mechanism	Contaminant	Media	Plant	Reference
Phytoextraction	Zn, Cd, and As	Soil	*Datura stramonium* L. and *Chenopodium murale* L.	Varun *et al.,* (2012)
Phytodegradation	Pb, Cd	Soil	*Jatropha curcas* L.	Mangkoedihardjo and Surahmadiada (2008)
Phytostabilisation	Cd	Soil	*Helianthus annuus* L.	Zadeh *et al.,* (2008)
Extraction-	Cd, Co, Cu,			
concentration in shoot	Ni, Pb and Zn	Wet lands	*Ipomoea saaquatica* Forsk, *Eichhornia crassipes* (Mart.) Solms, *Typha angustata* Bory and Chaub, *Echinochloa colona* (L.) Link, *Hydrilla verticillata* (L.f.) Royle, *Nelumbo nucifera* Gaerth and *Vallisneria spiralis* L.	Kumar *et al.,* (2008)
Phytodegradation	Total petroleum hydrocarbons (TPH)	Soil	*Anogeissus latifolia* (Roxb. ex DC.) Wall, *Terminalia arjuna* (Roxb.) Wight. *Tacomella undulata.* (Sm.) Seem.	Mathur *et al.,* (2010)
Phytodegradation	Zn and Cd	Soil	*Vetiveria, Sesbania, Viola, Sedum, Rumex.*	Mukhopadhyay and Maiti, (2010).
Phytodegradation	As	Soil	*Cassia fistula* Willd.	Preeti *et al.,* (2011)
Phytoextraction	Cr	Soil	*Anogeissus latifolia* (Roxb. ex DC.) Wall,	Mathur *et al.,* (2010)
Phytoextraction	137Cs	Soil	*Catharanthus roseus* (L.) G. Don	Fulekar *et al.,* (2010)
Phytodegradation	U	Soil	*Brassica juncea* (L.) Czern.	Huhle *et al.,* (2008)
Phytoextraction	Uranium and Thorium	Soil	*Nyssa sylvatica,* Marsh. *Liquidambar styraciflua* L.	Saritz (2005)
Phytoextraction	Mn	Soil	*Cousinia bijarensis* Rech.f. *Chondrila juncea* L. *Chenopodium botrys* (L.) Mosyakin & Clemants	Cheraghi *et al.,* (2011)

3
Plants as tool for Phytoremediation

3.1 Introduction

The principal application of phytoremediation is for lightly contaminated soils and waters where the material to be treated is at a shallow or medium depth and the area to be treated is large. This will make agronomic techniques economical and applicable for both planting and harvesting. In addition, the site owner must be prepared to accept a longer remediation period. Plants that are able to decontaminate soils do one or more of the following:

1. Plant uptake of contaminant from soil particles or soil liquid into their roots

2. Bind the contaminant into their root tissue, physically or chemically and

3. Transport the contaminant from their roots into growing shoots and prevent or inhibit the contaminant from leaching out of the soil.

Moreover, the plants should not only accumulate, degrade or volatilize the contaminants, but should also grow quickly in a range of different conditions and tend themselves to easy harvesting. If the plants are left to die *in situ*, the contaminants will return to the soil. So, for complete removal of contaminants from an area, the plants must be cut and disposed of elsewhere in a non-polluting way. Some examples of plants used in phyromediation practices are the following; water hyacinths (*Eichornia crassipes*); poplar trees (*Populus* sps.); forage kochia (*Kochia* sps); alfalfa (*Medicago sativa*); Kentucky bluegrass (*Poa protensis*); Scirpus spp, coontail (*Ceratophyllum demersum* L.); American pondweed (*Potamogeton nodosus*) and the emergent common arrowhead (*Sagittaria latifolia*) among others.

Four heavy metal concentrations to soils (Cu, Cr, As and Pb) were examined. Tomato and mustard plants were able to extract different concentrations of each heavy metal from the soils. The length of time that the soils were exposed to the contaminants affected the levels of heavy metals accumulation. Today, many institutions and companies are funding scientific efforts to test different plants effectiveness in removing wide ranges of contaminants. Scientists favor *Brassica juncea* and *Brassica olearacea*, two members of the mustard family, for phytoremediation because these plants appeared to remove large quantities of Cr, Pb, Cu and Ni from the soil.

3.2 Phytoremediation as a cleansing tool

It is actually a generic term for several ways in which plants can be used for these purposes. It is characterized by the use of vegetative species for *in situ* treatment of land areas polluted by a variety of hazardous substances (Sykes, Vina and Abubakr, 1999).

Garbisu (2002), have defined phytoremediation as an emerging cost effective, non-intrusive, aesthetically pleasing and low cost technology using the remarkable ability of plants to metabolize various elements and compounds from the environment in their tissues. Phytoremediation technology is applicable to a broad range of contaminants, including metals and radionuclides, as well as organic compounds like chlorinated solvents, polychloribiphenyls, polycyclic aromatic hydrocarbons, pesticides / insecticides, explosives and surfactants. According to Macek (2004), phytoremediation is the direct use of green plants to degrade, contain or render harmless various environmental contaminants, including recalcitrant organic compounds or heavy metals. Plants are especially useful in the process of bioremediation because they prevent erosion and leaching that can spread the toxic substances to surrounding areas.

Next process is the phytostabilization, which makes plants release certain chemicals that bind with the contaminant to make it less bioavailable and less mobile in the surrounding environment. Last is phytovolitalization, a process through which plants extract pollutants from the soil and then convert them into a gas that can be safely released into the atmosphere (Clemens, Palmgren and Kramer, 2002). Rhizofiltration is a similar concept to phytoextraction, but mainly use with the remediation of contaminated groundwater rather than the remediation of polluted soils. The contaminants are either absorbed onto the root surface or are absorb by the plant roots. Plants used for rhizofiltration are not planted directly *in situ*, but are acclimated with the pollutant first. Until a large root system has developed, plants are hydroponically grown in clean water rather than in soil.

Research has been slow and tedious due to scientist's incomplete understanding of the generalized cellular mechanisms of plants. However, the advent of new genetic technology has allowed scientists to determine the genetic basis for high rates of accumulation of toxic substances in plants (Moffat, 1995). Using genetic engineering, scientists may soon be able to exploit plants characteristics that can provide faster and more efficient means of removing contaminants from the soil. Genetic engineering will also be crucial for the creation of transgenic plants that will be able to combine the natural agronomic benefits associated with plants (ease of harvest and rapid, expansive growth) with the remediation capabilities of bacteria-a traditional organism used in bioremediation (Huang and Cunningham, 1996).

Phytoremediation of heavy metals from the environment serves as an excellent example of plant-facilitated bioremediation process and its role in removing environmental stress. Traditionally, when an area becomes contaminated with heavy metals, the area must be excavated and the soil should be removed and put to a landfill site. This process is extremely expensive and, therefore not entirely appealing despite recent discoveries regarding phytoremediation (Lasat, 2000).

Therefore, scientists have been forced to become very creative in developing effective transgenic phytoremediators.

Many human diseases result from the builtup of toxic metals in soil, making remediation crucial in protecting human health. Lead is one of the most difficult contaminant to be removed from the soil and one of the most dangerous. According to Lasat (2000), the presence of Pb in the environment can have devastating effects on plant growth and can result in serious side effects - including seizures and mental retardation if ingested by humans or animals. Much of the global Pb contamination has occurred as a result of mining and iron smelting activities (Cortez, 2005). Phytoremediation of Pb contaminated soil involves two of the aforementioned strategies – phytostabilization and phytoextraction. It is believed that plants ability to phytoextract certain metal is a result of its dependence upon the absorption of many metals such as Zn, Mn, Ni, and Cu (Lasat, 2000).

3.3 Phytoremediation in curing soil problems with crops

Currently, conventional remediation methods of Heavy Metals (HM) contaminated soils are expensive and environmentally destructive (Bio-Wise, 2003; Aboulroos *et al.*, 2006). Since then, scientists all over have been in search of some innovative, eco-friendly and low cost alternative technologies. One of them is the phytoremediation, which include the use of plants to clean and cure the environment; and plants have been known for their property to absorb, accumulate and detoxify the impurities present in the soil, water and air through various physical, chemical and biological processes (Hooda, 2007). Phytoremediation, a fast-emerging new technology for removal of toxic HMs, is cost-effective, non-intrusive and aesthetically pleasing. It exploits the ability of selected plants to remediate pollutants from contaminated sites. Plants have inter-linked physiological and molecular mechanisms of tolerance of HMs. High tolerance to HM toxicity is based on a reduced metal uptake or increased internal sequestration, which is manifested by interaction between a genotype and its environment. The growing interest in molecular genetics has increased our understanding of mechanisms of HM tolerance in plants and many transgenic plants have displayed increased HM tolerance. Improvement of plants by genetic engineering, that is, by modifying characteristics like metal uptake, transport and accumulation and plant's tolerance to metals, open up new possibilities of phytoremediation. Either naturally occurring or genetically engineered plants are used for cleaning contaminated environments. Phytoremediation can be used to remove not only metals (for example, Ag, Cd, Co, Cr, Cu, Hg, Mn, Mo, Ni, Pb, Zn) but also redionuclides (for examples, ^{90}Sr, ^{137}Cs, ^{239}Pu, ^{234}U, ^{238}U) and certain organic compounds (Andrade and Mahler, 2002). The phyloremediation efficiency of plants depends upon various physical and chemical properties of soil, plant, bioavailability of metals and capacity of plants to uptake, accumulate and detoxify metals. For selections of plants which are suitable for phytoremediation of polluted soils, one has to understand the mechanism underlying plant tolerance towards a particular metal. The HM pollution is a very vast subject, but in this review, we will try to focus on the sources of soil pollution, mechanism of metal uptake by the plants and the different types of phytoremediation and their practical application in soil remediation.

3.4 Grasses as Potential phytoremediators Vetiver Grass (*Vetiveria zizaniodes L.*)

Vetiver (*Vetiveria zizanioides* L.), this plant belongs to the same grass family as maize, sorghum, sugarcane and lemon grass. It has several unique characteristic as reported by the National Research Council. Vetiver grass is a perennial grass growing two meters high, and three meters deep to the ground. It has a strong dense and vertical root system. It grows both in hydrophytic and xerophytic conditions. The leaves sprout from the bottom of the clumps and each blade is narrow, long and coarse. The leaf is 45-100 long and 6-12 cm wide.

Vetiver grass is highly suitable for phytoremedial application due to its extraordinary features. These include a massive and deep root system, tolerance to extreme climatic variations such as prolonged drought, flood, submergence, fire, frost and heat waves. It is also tolerant to a wide range of soil acidity, alkalinity, salinity, sodicity, elevated levels of Al, Mn and heavy metals such as As, Cr, Ni, Pb, Zn, Hg, Se and Cu in soils (Troung and Baker, 1996).

In Thailand, Vetiver grass is found widely distributed naturally in all parts of the country. It has been used for erosion control and slope stabilization. Vetiver hedges had an important role in the process of captivity and decontamination of pesticides, preventing them from contaminating and accumulating in crops (Troung, 1995). When compared with other plants, Vetiver grass is more efficient in absorbing certain heavy metals and chemicals due to the capacity of its root system to reach greater depths and widths (Troung and Baker, 1996). As confirmed by Roongtanakiat and Chairoj (2010). Vetiver grass was found to be highly tolerant to an extremely adverse condition. Therefore, Vetiver grass can be used for rehabilitation of mine tailings, garbage landfills and industrial waste dumps which are often extremely acidic or alkaline, high in heavy metals and low in plant nutrients.

Cogon Grass (*Imperata cylindrica L.*)

Cogon grass, generally occurs on light textured acid soils with clay subsoil, and can tolerate a wide range of soil pH ranging from strongly acidic to slightly alkaline (Johnson and Shilling, 2001). It is hardy species, tolerant of shade, high salinity, and drought. It can be found in virtually any ecosystem, especially those experiencing disturbances (Mekonnen, 2000). It is a perennial grass up to 120 cm high with narrow and rigid leaf-blades. The roots can penetrate to a soil depth of about 58cm to alluvial soil. More than 80 percent of shoots can originate from rhizomes less than 15 cm below the soil surface.

Carabao Grass (*Paspalum conjugatum L.*)

Carabao grass is a vigorous, creeping perennial grass with long stolons and rooting at nodes. Its culms can ascend to about 40 to 100 cm tall, branching, solid and slightly compressed where new shoots can develop at every rooted node. Under a coconut plantation, a yield of about 19000 kg ha^{-1} of green materials was obtained. It grows from near sea-level up to 1700 m altitude in open to moderately shaded places. It is adapted to humid eliminates and found growing gregariously under plantation crops and also along stream banks, roadsides, and in disturbed areas. This grass can adapt easily to a wide range of soils (Mannetje, 2004).

3.5 Phytoremediation from Saline tolerant Plants

In arid land when the amount of precipitation is limited high evaporation rates and slow drainage allow salt accumulation and concentration in the soil, thus rendering these soils unproductive. Salt problems in different soils can occur under almost any climatic region. Once the salinity level exceeds the common tolerable limits of growth by plants, the land becomes fallow and unusable agricultural resource.

Agricultural expansion in arid and semi-arid regions entails a great amount of irrigation water. Fresh water resources are not sufficient to be used in cultivation to meet and cover the food demand of an increasing population. Thus, exploring the possibility of using saline water for irrigation, especially drainage and underground water is of great importance. The applicability of saline water for irrigation is dependent upon the concentration, composition of salts dissolved therein and the degree to which plants can tolerate salt. Today, new appreciation of plant physiology, soil sciences and new irrigation techniques have shown that with careful management, saline water can be used to grow a variety of crops and more desert and could be cultivated with different economic plants as reported in Saline Agriculture (1990).

A general reduction in growth and yield attributes in response to salinity is widely documented. The major factor is mainly due to limited supply of metabolic energy for maintenance of normal growth processes. Salinity increases the amount of work necessary to counteract osmotic and ionic stresses for normal cellular maintenance as a consequence, there is less energy left for growth requirements (Greenway, 1962; Stayter, 1967; El-Saidi and Hawash, 1971; El-Shaidi, 1973; El-Saidi and Kortam, 1974; El-Saidi *et al.*, 1983; El-Shadi *et al.*, 1986; Kramer, 1983; Austin, 1989 and Kozlowski *et al.*, 1991).

Fodder beet

Growing *Beta vulgaris*, important cattle feed as a salt tolerant crop in saline soils, or using saline water for irrigation of this crop has important economic value, especially in sandy soils. This crop can grow successfully under drop irrigation in newly reclaimed soils and gives high economic yield of roots. The rapid infiltration of water through sandy soil reduces the salt buildup in the root zone. It is recommended to grow fodder beet by transplanting the seedling previously grown in the nursery. This gives better results concerning the yield of roots. Direct seed sowing could delay germination and seed emergence.

Fodder beet is sensitive to salinity at early seedling stage and acquires tolerance with age. The tolerance of salinity varies and is mainly related to the plant type and its growing stage, as well as the levels of salts in soil or in irrigation water. Growth of fodder beet under drop irrigation was satisfactory and gave good economical yield. However, care must be taken in land preparation and planting methods to control salinity. Irrigation procedures by droppers can maintain a relatively high soil moisture regime and periodically could leach accumulated salts in the soil. Growing fodder beet under drip irrigation system gives high yield of leaves and roots. Some plants gave root weight of about 40-45 kg/plant (El-Saidi and Ali, 1993).

Cotton Plants

Trials were conducted to grow cotton plants in sandy saline soil which contain about 15 per cent of calcium carbonate. Salts present in the seed bed reduced the rate of germination to about 10 per cent. The time of emergences became longer. To circumvent these effects of salinity, cotton seeds were presoaked for about 24-36 hours before sowing to activate the seed embryo. The appearance of crust layers on the upper part of the soil surface (lines) over the hills disturbed the growth of seedling. This also increased the opportunity for disease problem to develop due to the delay in emergence.

Feasibility of using saline water for irrigating cotton plants is determined by its salt content and the developmental stage of cotton plant, during which saline water applied (El-Saidi *et al.*, 1983, 1986, 1992). They found that the most sensitive stage to salinity in cotton plants was the squaring salinity affects the growth of cotton plants by completely stopping the growth and higher rate of shedding takes place. Consequently, reduction in seed cotton yield was very severe (about 80 per cent) in comparison with normal irrigated plants. The average weights of bolls and seeds were reduced.

Rice (*Oryza sativa*)

Rice (*Oryza sativa*) is moderately sensitive to salinity. The salinity threshold effect is close to ECe 3 ds/m, beyond which yield starts to decrease. Salinity affects almost all phases of growth of the rice plants and decreases the yield. A decrease in photosynthetic activity was observed and chlorophyll content obviously reduced at higher levels of salinity. Salinity affects all growth parameters, plant heights, root length, number of tillers per plant, straw fresh and dry weight. Under salinity conditions panicle length is severely reduced, along with the number of primary branches per panicle and seed set percentage and panicle weight and thus a reduction in the grain yield are evident.

Barley

Barley (*Hordeum vulgare* L.) is the most important field crop in dry areas, especially in countries of West Asia and North Africa. However, a big gap exists between the yield potential and the actual production. Barley can grow in different zones in dry area, where it is of major significance to the livelihood of millions of rural families. Barley can resist drought. It can grow and yield economically under severe conditions of water deficit. It is the most useful grain crop for Beduins living in the desert. They can use it in making bread and/or a fodder for their animals (camels, sheep, and goats). It is also the most salt tolerant cereal grain plant.

3.6 List of plants used in phytoremediation

A list of plants successfully utilized for phytoremediation studies is given in the below Table.5.

Table 5: List of Plants used in Phytoremediation of Organic and Inorganic Pollutants

	Scientific Name	Common Name
1.	*Achillea millefolium* L.	european milfoil; yarrow
2.	*Agropyron smithii* Rydb.	prairie grass
3.	*Agropyron trachycaulum* (Link)Make	slender wheatgrass
4.	*Agrostis tenuis* Linn.	colonial bent grass
5.	*Atriplex canescens* (Pursh) Nutt.	four-wing saltbrush
6.	*Alyssum wulfenianum* Bernh.	-
7.	*Andropogon gerardii* Vitaman.	big blue stem
8.	*Andropogon scoparius* Michx.	little blue stem prairie grass
9.	*Arabidopsis thaliana* (L.) Heynh.	mustard weed
10.	*Armeria maritima* (Mill.) Willd.	sea pink; wild thrift
11.	*Armoracia rusticana* P. Gaertn.	horse radish
12.	*Astragalus racemosus* Pursh.	-
13.	*Azolla pinnata* R. Br.	Water velvet
14.	*Azolla filiculoides* Lam.	-
15.	*Betula nigra* L.	river birch
16.	*Brassica juncea* L.	Indian mustard
17.	*Brassica napus* (L.) Czern	canola
18.	*Buchloe dactyloides* (Nutt.) Engelm.	buffalo grass
19.	*Chenopodium album* L.	Goose foot
20.	*Cynodon dactylon* (L.) Pers.	bermuda grass
21.	*Datura inoxia* Mill.	Thornapple
22.	*Dicoma niccolifera* Wild.	-
23.	*Eichhornia crassipes* (Mart.) Solms	Water hyacinth
24.	*Elymus canadensis* L.	Canada wild rye
25.	*Festuca arundinacea* Schreb.	Tall fescue
26.	*Festuca ovina* L.	hard fescue
27.	*Festuca rubra* L.	red fescue
28.	*Geissois pruinosa* Brongn. & Gris	-
29.	*Glycine max* (L.) Merr.	soybean
30.	*Haumaniastrum robertii* (Robyns) P.A.Duvign	-
31.	*Helianthus annuus* L.	sunflower
32.	*Hibiscus cannabinus* L.	Indian kenaf
33.	*Hybanthus floribundus* (Lindl.) F.Muell	Green violet
34.	*Hydrocotyle umbellata* L.	pennywort

Scientific Name	Common Name
35. *Ipomoea* sp.	-
36. *Lemna minor* L.	Duckweed
37. *Lespedeza cuneata* (Dum, Cours.) G.Don	Chinese bush-clover
38. *Leucaena leucocephala* (Lam.) de Wit	Koa haole (White Lead Tree)
39. *Liriodendron tulipifera* L.	Yellow poplar
40. *Maclura pomifera* (Raf.) C.K.Schneid	Osage orange
41. *Malus fusca* (Raf.) C.K.Schneid	Crab apple
42. *Medicago sativa* L.	Alfalfa
43. *Mentha spicata* L.	Spear mint
44. *Morus rubra* L.	Red mulberry
45. *Myriophyllum aquaticum* (Vell.) Verdc.	Parrot feather
46. *Myriophyllum spicatum* L.	Water milfoil
47. *Nicotiana tobacum* L.	Tobacco
48. *Nitella flexilis* (L.) C.Agardh	Stonewort
49. *Oryza sativa* L.	Rice
50. *Oryza sativa* subsp *indica*	Indian ricegrass
51. *Panicum miliaceum* L.	Proso millet
52. *Panicum virgatum* L.	Switchgrass (Prairie grass)
53. *Pearsonia metallifera* Wild.	-
54. *Phaseolus coccineous* L.	Runner bean
55. *Phaseolus vulgaris* L.	Kidney bean
56. *Phragmites australis* (Cav.) Trin.ex Steud	Reed grass
57. *Pinus taeda* L.	Loblolly pine
58. *Polygonum hydropiperoides* Michx.	Swamp smart weed
59. *Populus angustifolia* James	Cottonwood
60. *Populus charkowiieensisxintrcrassata*	Hybrid poplar
61. *Populus deltoids* W.Bartram ex Marshall	Eastern cotton wood
62. *Populus tricocarpa* Torr.& A.Gray ex Hook	Black cotton wood
63. *Prosopis julifera* (Sw.) DC	mesquite
64. *Psychotria douarrei* (G. Beauvisage) Daniker	cherry bark oak
65. *Quercus falcata* Michx.	southern red oak
66. *Quercus virginiana* Mill.	live oak
67. *Robinia pseudoacacia* L.	black locust
68. *Rinorea bengalensis* (Wall.) Kuntze	-
69. *Rinorea javanica* (Blume) Kuntze	-
70. *Sebertia acuminate* Pierre ex Baill	-
71. *Saccharum officinarum* L.	sugarcane
72. *Salix alaxensis* (Anderson) Coville	felt leaf willow
73. *Salix nigra* Marsh.	black willow tree
74. *Salvinia molesta* D.S.Mitch	kariba weed

Scientific Name	Common Name
75. *Silene vulgaris* L.	bladder campion
76. *Sorghastrum nutans* (L.) Nash	Indian grass
77. *Sorghum vulgare* Hitchcock	Sudan grass
78. *Spartina alterniflora* Loisel	salt marsh cord grass
79. *Taxodium distichum* (L.) Rich.	bald cypress
80. *Thlaspi caerulescens* J.& C. Presl	alpine pennycress
81. *Thalaspi rotundifolium* (L.) Gaudin	Pennycress
82. *Thlaspi goesingense* Halacsy	Tiny wild mustard
83. *Typha latifolia* L.	Cat tail
84. *Typha orientalis* (C.Presl) Rohrbach	Bull rush

One of the most debated controversies in the field refers to the choice of remediative species; metal hyperaccumulators vs. common nonaccumulator species. Since the rate of metal removal depends upon the quantity of biomass and metal concentration in the biomass per harvest. Hyper-accumulator plants have the potential to bioconcentrate high metal levels; however their use may be limited by small size and slow growth. For vigorous Phyto-extraction of metals the design requirements are; vegetation should be fast growing and tolerant, easy to plant and maintain. In common nonaccumulator species, low potential for metal bioconcentration is often compensated by the production of significant biomass (Ebbs and Kochain, 1997). Chaney *et al.*, (1999), did not support this argument but his study was limited to Zn and Cd removal, with few non-accumulator crops. In support of this, several maize (one of the most productive crops) inbred lines have been identified which can accumulate high levels of Cd (Hinesly *et al.*, 1978). However, these lines were susceptible to Zn toxicity and, therefore, could not be used to cleanup soils at the normal Zn: Cd ratio of 100:1 (Chaney *et al.*, 1999). Plants showing high Phyto-extraction include Chenopodium leaves. It had relatively higher arsenic concentrations (14 mg/kg) than other native plant or poplar leaves (8 mg/kg) in mine-tailing wastes (Pierzynski *et. al.*, 1994). *Helianthus annuus* took up Cs and Sr with Cs remaining in the roots and Sr moving into the shoots (Adler, 1996). Metal accumulator plants such as the crop plants corn, sorghum, and alfalfa may be more effective than hyperaccumulators and remove a greater mass of metals due to their faster growth rate and larger biomass. Baker (1995) has found 80 species of nickel-accumulating plants in the Buxaceae (including boxwood) and Euphoribiaceae (including catctus-liks succulents) families some euphorbs can accumulate up to 5% of their dry weight in nickel.

Phytoremediation techniques like rhizofiltration are limited to remediate aqueous pollutants need large constructions and are a slow process with quite low efficiency. Phyto-volatilization is only possible with mercury and selenium, as they can be volatilized in contrast Phyto-extraction works for removal of pollutant permanently and require the availability of metals to plants. Broadening the applicability will require using plants that perform better and have high biomass production, heavy metal uptake and translocation capacity. Ecological studies have revealed the existence of specific plant communities, endemic plants, which have adapted on soils contaminated with elevated levels of Zn Cu, and Ni. Different ecotypes of the same species may occur in areas uncontaminated by metals. To plants endemic to metal-contaminated soils, metal tolerance is an indispensable property. In comparison, in related populations inhabiting uncontaminated areas, a continuous gradation between ecotypes with high and low tolerance usually occurs.

Biological Treatment Technologies

4.1 Plants for Bioremediation of Contaminated Land

Polluted soil remediation technologies follow two main pathways 'concentrate and isolate' or 'dilute and disperse', the pollutant below the threshold concentration. Contaminated soil can be remediated by physical, chemical and biological techniques; all these available techniques may be grouped into two categories, *ex-situ* and *in-situ* (Baker and Walker, 1990). For Some pollutants and/or sites the technique applied can be both *in-situ* and *ex-situ*. *Ex-situ* method requires removal of contaminated soil for treatment on or of site, and returning the treated soil to the restored site. It relies on excavation, detoxification and /or chemically. In situ method is remediation without excavation of contaminated site. Reed *et al.*, (1992), have defined *in-situ* remediation technologies as destruction or transformation of the contaminant, immobilization to reduce bioavailability and separation of the contaminant from the bulk soil. *In-situ* techniques are favored over the *ex-situ* techniques due to their low cost and reduced impact on the ecosystem. This chapter is limited to biological treatment technologies and elaborates on the requirements and types of plants reported for successful phyto-remediation or soil pollutants.

The Biological technologies were commonly used for the remediation of organic contaminants but presently are being applied for metal remediation too (Schnoor, 1997). Simple bioremediation involves the uptake of metals from contaminated media by living organisms or dead, immobilization biomass reported by (Lakshman, 1998). Adsorption to ionic groups on the cell surface is the primary mechanism for metal adsorption by inactive biomass (Means and Hinchee, 1994). Active plants and microorganisms accumulate metals as the result of normal metabolic processes via ion exchange at the cell walls or intra and extracellular precipitation and complicated reactions; all these have been classified under Bioremediation.

Growth of plants depends on photosynthesis, in which water and carbon dioxide are converted into carbohydrates and oxygen, using the energy from sunlight. Similarly, plants growing in sewage polluted soil with AM fungal colonization are needed to be examined (Lakshman, 2000; Lakshman and Hosamani, 2003). The idea of using metal accumulating plants to remove heavy metal and other compounds was first introduced in 1983, but the concept has actually been implemented for the past 300 years (Henry, 2000).

For Phytoremediation of contaminated site it is necessary to carry out some pre-plant selection activities which include gathering site information on:

1. Site location, contaminant species, contaminant concentration, soil (texture, structure, pH, salinity, fertility, water content, leaching properties) and depth of bedrock.

2. Solar radiation, annual rainfall, irrigation requirements and other infrastructure.

3. Soil must have a pH range that will allow plant growth. The soil pH might need to be modified and controlled through liming to increase pH or through acidification to lower pH.

4. Identify appropriate phytoremediation technology and remedial limits as per the law of the land.

4.2 Plant Selection activities for Phytoremediation

The selection of plant species is possibly the single most important factor affecting the extent of metal removal. Use of plants adds an additional level of complexity to the remediation process because plants comprise a complex biological system that has its own characteristics. As plants are the primary producers, they are directly responsible for any contamination of the food chain; hence professionals involved in situ remediation should be careful about the selection of the plants. Although, the potential for pollutant extraction is of primary importance, other criteria, such as ecosystem protection must be also considered when selecting plants. The primary requirement is that the plants should be able to grow vigorously in contaminated soils. Therefore plants should be selected according to the need of application and the contaminants of concern. Site-specific conditions or considerations like identification of local plants and crops growing in contaminated area can provide a clue, as to what plants to select. The agronomic considerations/ factors that are considered in successful agriculture also must be considered during phytoremediation. These factors will be more critical or more difficult to control due to the additional stress placed on the system by the contamination. As a general rule, native species are preferred to exotic plants, as exotic species can be invasive and endanger the harmony of the ecosystem.

The goal of the plant selection process is to choose a plant species with appropriate characteristics for growth under site conditions that meet the objectives of phytoremediation. The points that should be taken into consideration for phytoremediation are:

1. Ploughing the land evenly distributes the contaminant throughout the field and also reduces the concentration to some level. A surface layer with minimal contamination underlain by greater contaminant concentrations might allow more successful seed germination than if the surface layer is heavily contaminated. The contaminant concentrations level should not be phytotoxic found at the site; this is to ensure that the plants will be able to grow initially. Contaminant phytotoxicity and uptake information can be found in the phytoremediation and agricultural literature, or determined through preliminary phytotoxicity screening and germination studies.

2. Conduct preliminary studies to assess germination and survivability of the plants. Germination and screening studies for phytotoxicity can be done in small-scale greenhouse, laboratory chamber studies and field pot trials. Since phytoremidiation can be a long term process, preliminary trials won't substantially increase the overall remediation time. Plants should be disease resistant, insect tolerant, drought resistant, salt tolerant and stress tolerant annual, biennial or perennial. Plant or seed source, seed pretreatment before germination (if needed) and planting method of seeds, plugs, transplant, their timing, density, depth should be taken care of before starting.

3. The plant can take up and degrade contaminants, produce exudates that can stimulate the soil microbes, or possess enzymes that are known to degrade a contaminant. The potential for the success of phytoremediation can be increased by screening plants for useful enzymes (Fletcher *et al.*, 1995). Monitor and evaluate plant growth and phytoremediation success, screening of cultivers or varieties might be required.

4. Establishment rate, reproduction rate, growth rate, biomass production, competitive or allelopathic effects, irrigation, fertilization, protection from pests and disease, spread of fallen leaves, debris and harvesting requirements should also be taken care of. The use of mixed species of vegetation can lead to more success due to the increased chance that at least on species will find a niche. A monoculture relies on just one plant type, possibly requiring less management; stand of one plant that has been shown to be effective could be the most efficient means of phytoremediation.

5. Native, nonagricultural plants are desirable for ecosystem restoration, in most applications; plants that are adapted to local conditions will have more chance of success than non adapted plants. Landfill or polluted sites though remote from human activities are always deficient in irrigation facilities, but they are always open for the movement of faunas. The animals utilize available food material from plant biomass, thus increasing the chances of introducing the heavy metal/pollutant into the food chain. To avoid metal being transferred to food chain; harmless weedy species in place of crops, should be preferred in general. Since most crops are palatable and pose risk to grazing animals and ultimately to humans.

6. Aquatic plants should also be considered; mostly the emergent species and species which can grow in marshy conditions have been studied, as they

can successfully grow along the effluent streams. These sites are neither managed nor are the conditions proper to facilitate plant all these limitations should be taken into consideration, which can species like water hyacinth, which can grow in deficient conditions, have good biomass potential, low to moderate uptake and translocation capacity and probably have high regeneration capacity.

7. Root contact is a primary limitation of phytoremediation applicability. Remediation with plants requires that the contaminants be in contact with the root zone of the plants. The range of root depths of a given plant must be considered according to the contaminant depth and distribution. The distribution of the contamination at various soil depths is also important in planning the plant type and planting method. The contaminant concentrations in the seedbed layer of the soil profile may have a strong effect on the ability to establish vegetation. Either the plants must be able to extend roots to the contaminants, or the contaminated media must be moved to within range of the plants. This movement can be accomplished with standard agricultural practices, such as deep ploughing to bring soil from two feet deep to within 20-30 cm of the surface for the shallow-rooted species would be appropriate to use, as most of the metal contaminants remain adsorbed in the upper layer of up to 6-8 cm. whereas deep-rooted plants like populous would be applicable for more profound contamination like volatile organic compound contamination of groundwater and aquifers (Ruskin, 2000). The root depth ranges provided below represent maximum depths: Some grass fibrous root systems can extend 8 to 10 feet deep. The roots of major prairie grasses can extend to about 6 to 10 feet. Shrub: The roots of Phreatophytic shrubs can extend to about 20 feet (Woodward, 1996). Trees: Phreatophytic tree roots can be as deep as 80 feet. Example is mesquite taproots, which range from 40 to 100 feet and river birch tap roots which go to 90 to 100 feet (Woodward 1996). *Alfalfa* a type of legume roots can go quite deep, down to about 30 feet, given the proper conditions. Other plants like Indian mustard roots generally are about 6 to 9 inches deep. These maximum depths are not likely to be reached in most situations, due to typical site conditions such as soil moisture being available in the surface soils or poorer soil conditions at greater depths. A review of the literature found that maximum depths of tree roots were generally 3 to 6 feet, with almost 90% o the roots in the top 2 feet (Dobson and Moffat, 1995). A fibrous root system has numerous fine roots spread throughout the soil and will provide maximum contact with the soil due to the high surface area of the roots. Fescue is an example of a plant with a fibrous root system. A tap root system is dominated by a central root, *Medicago sativa* is an example of a plant with a tap root system was studied by (Schwab *et. al.*, 1998).

8. Plant Growth - Growth is the change in size over time, plant growth indicates its ability to utilize resources. The growth rate of a plant will directly affect the rate of remediation measurement of plant growth can be useful for comparing plants. Growth rates can be defined differently for

different forms of phytoremediation. And phytostabilization, it is desirable to have fast growth in terms of root depth, density, volume, surface area, and lateral extension. For phytoextraction, a fast growth rate of above ground plant mass is desirable. A large root mass and large biomass is desired for an increased mass of accumulated contaminants, for greater transpiration of water, greater assimilation and metabolism of a greater amount of exudates and enzymes. A fast growth rate will minimize the time required to reach a large biomass.

9. Plant rotation could be important for metal remediation when short-lived vegetation is used that does not reach remedial goals and that should not be replanted in the same place. The long term establishment of vegetation at a site is dependent on the project goals and future uses of a site. For long term, no-maintenance vegetation establishment as part of ecosystem restoration, it is likely that there will be a succession of plants at a site. If so, this succession could be planned when considering the types and timing of vegetation. It is more applicable in Phytostabilization of contaminated site.

10. Metal hyper accumulators are able to concentrate a very high level of some metals, however their generally low biomass and slow growth rate means that the total mass of metals removed will tend to be low. For phyto-extraction of metals, the metals concentration in the biomass and the amount of biomass produced must both be considered. A plant that extracts a lower concentration of metals, but that has a much greater biomass than many hyper accumulators, is more desirable than the hyper-accumulator because the total mass of metals removed will be greater. Poplars have been widely used in phytoremediation research and applications due to their fast growth rate; they can grow 9 to 15 feet/year reported by (Gordon, 1997).

11. The transpiration rate of vegetation will be important for those phytoremediation technologies that involve contaminant uptake, and for hydraulic control. The transpiration rate depends on factors such as species, age, mass, size, leaf surface area, canopy cover, growth stage, and climatic factors, and will vary seasonally. Thus, well-defined numbers for a given a type of vegetation cannot be assumed. Estimates for certain cases, however, provide a rough guide to the order if magnitude that might be expected, for Populous species; approximately 26 gpd (gallons per day) for a 5 years old tree was estimated by (Gordon, 1997). It was reported that 5000 gpd was transpired by a single willow tree, which is comparable to the transpiration rate of 0.6 acre of *Alfalfa*. Individual cottonwood trees were estimated to transpire between 50 and 350 gpd, based on analysis of drawdown near the trees recorded by (Gatliff 1994). These transpiration rates, given in terms of gpd for individual trees, should be viewed with caution because the transpiration rate varies with tree size and other factors. A more appropriate measure would be to look at the total water usage in a given area of vegetation.

5

Phytoremediation of Environmental pollutants

5.1 Phytoremediation of Radioactively contaminated sites

The application of nuclear energy and the use of radionuclides for industrial, medical and research purposes have caused significant contamination of certain sites and their environment, which could result in health problems for several centuries if nothing is undertaken to remedy these situations. Except for the immediate environment of the facility, where decontamination activities may be feasible and affordable, the contamination often extents over a vast area and decontamination would be costly and could result in vast amounts of waste. Therefore, more realistic yet efficient remediation options should be searched for of which phytomanagement is among the potential options.

Phytoremediation of [137]cesium and [90]strontium from solutions and low-level nuclear waste by *Vetiveria zizanoides.* L.

Vetiver grass (*Vetiveria zizanoides* L. Nash) plantlets when tested for their potential to remove ^{90}Sr and [137]Cs (5 10³ k Bq 1¹) from solutions spiked with individual radionuclide showed that 94% of ^{90}Sr and 61% of ^{137}Cs could be removed from solutions after 168 h. When both ^{90}Sr and ^{137}Cs were supplemented together to the solution, 91% of ^{90}Sr and 59% of ^{137}Cs were removed at the end of 168 h. In case of ^{137}Cs, accumulation occurred more in roots than shoots, while ^{90}Sr accumulated more in shoots than roots. When experiments were performed (Singh *et. al.*, 2008) to study the effect of analogous elements, K^+ ions reduced the uptake of ^{137}Cs, while ^{90}Sr accumulation was found to decrease in the presence of Ca^{2+} ions. Plants of *V. zizanoides* could also effectively remove radioactive elements from low-level nuclear waste and the level of radioactivity was reduced below detection limit at the end of 15 days of exposure. Singh *et al.*, (2008) suggestested that *V. zizanoides* may be considered as a potential candidate plant for phytoremediation of ^{90}Sr and [137]Cs.

Phytoremediation of radiostrontium (^{90}Sr) and radiocesium (^{137}Cs) using giant milky weed (*Calotropis gigantea* R.Br.) plants

Potential of plants to remove radionuclides/toxic elements from soils and solutions can be successfully applied for removal of important radionuclides such as strontium-90 (^{90}Sr) and cesium-137 (^{137}Cs). When uptake of ^{137}Cs and ^{90}Sr by *Calotropis gigantea* plants incubated in distilled water spiked with the radionuclides either alone or in combination was studied (Eapen *et. al.*, 2006), it was found to have a high efficiency for the removal of ^{90}Sr, with 90% being removed from solutions (5 10^3 kBq l^1) within 24 h of incubation. However, in case of^{137}Cs, about 44% could be removed from solutions (5 10^3 k Bq l^1) at the end of 168 h of incubation. Accumulation of ^{90}Sr and ^{137}Cs was higher in roots compared to shoots. The plants could remediate both^{90}Sr and ^{137}Cs when they were added together to the solution. When two months old plants were incubated in low level nuclear waste, 99% of activity disappeared at the end of 15 days. Eapen *et. al.*, (2006) have suggested that *C. gigantea* may be used as a potential candidate plant for phytoremediation of ^{90}Sr and ^{137}Cs.

Phytoremediation of Uranium-Contaminated Soils

Uranium phytoextraction, the use of plants to extract U from contaminated soils, is an emerging technology. Jianweiw *et al.*, (1998) had reported on the development of this technology for the cleanup of U-contaminated soils. They investigated the effects of various soil amendments on U desorption from soil to soil solution, studied the physiological characteristics of U uptake and accumulation inplants, and developed techniques to trigger U hyperaccumulation in plants. A key to the success of U phytoextraction is to increase soil U availability to plants. They have found that some organic acids added to soils have increased U desorption from soil to soil solution and triggerered a rapid U accumulation in plants. Of the organic acids (acetic acid, citric acid, and malic acid) tested, citric acid was the most effective in enhancing U accumulation in plants. Shoot U concentrations of *Brassica juncea* and *Brassica chinensis* grown in a U-contaminated soil (total soil U, 750 mg kg-1) increased from less than 5 mg kg-1 to more than 5000 mg kg-1 in citric acid-treated soils. According to their knowledge, that was the highest shoot U concentration reported for plants grown on U-contaminated soils. Using this U hyperaccumulation technique, they were able to increase U accumulation in shoots of selected plant species grown in two U-contaminated soils (total soil U, 280 and 750 mg kg-1) by more than 1000-fold within a few days. Their results suggested that U phytoextraction may provide an environmentally friendly alternative for the cleanup of U-contaminated soils.

It is important to note that citric acid-triggered U hyperaccumulation in plants is rapid. The induced U hyperaccumulation in shoots can be observed within 24 h after the application of citric acid, and shoot U concentration reaches a maximum in 3 days. In this form of U phytoextraction, the plant would contain very low U concentration for most of its life. After the application of citric acid, U accumulation in plant shoots would increase rapidly, and the plants could be harvested a few days after the citric acid application. Using this technique, they were now able to increase U accumulation in shoots of selected plant species by more than 1000-fold within a few days. This strategy has advantages in reducing the risk that might have been present by having plants with high U levels in the field for long periods of time. Applying this technique in the field will speed up the removal of U from contaminated soils and provide a cost-effective soil decontamination strategy. Most importantly, this technology represents an environmentally friendly alternative for the cleanup of U-contaminated soils.

Water hyacinth for phytoremediation of radioactive waste simulate contaminated with cesium and cobalt radionuclides

Phytoremediation is based on the capability of plants to remove hazardous contaminants present in the environment. Saleh (2012) has studied some factors controlling the phytoremediation efficiency of live floating plant, water hyacinth (*Eichhornia crassipes*), towards the effluents contaminated with ^{137}Cs and/or[60]Co. Cesium has unknown vital biological role for plant while cobalt is one of the essential trace elements required for plant. He used undesirable species, water hyacinth, in purification of radiocontaminated aqueous solutions. The controlling factors such as radioactivity concentration, pH values, the amount of biomass and the light were studied. The uptake rate of radiocesium from the simulated waste solution was found inversely proportional to the initial activity content and directly proportional to the increase in mass of plant and sunlight exposure. A spiked solution of pH 4.9 was found to be the suitable medium for the treatment process. The uptake efficiency of ^{137}Cs present with ^{60}Co in mixed solution was higher than if it was present separately. On the contrary, uptake of ^{60}Co was affected negatively by the presence of ^{137}Cs in their mixed solution. Sunlight is the most required factor for the plant vitality and radiation resistance. Experimental results were proved that the water hyacinth as potential candidate plant of high concentration ratios (CR) for phytoremediation of radionuclides such as[137]Cs and [60]Co.

5.2 Phytoremediation of explosives

Contamination of soil and groundwater by explosive wastes and their transformation products is a common problem at explosive manufacturing/ processing, storage and disposal facilities around the world and need to be mitigated (Cataldo, D. A. 1990). Of nearly twenty different energetic compounds used in conventional munitions by the military today, hexahydro-1,3,5 trinitro-1,3,5 triazine (RDX) and octahydro-1,3,5,7- tetranitro-1,3,5,7- tetrazocine (HMX) are the most powerful and commonly used. These compounds are highly toxic (Cataldo, D.A. 1990) in nature and have resulted in severe soil and groundwater contamination due to their release in the environment.

Various treatment methods viz., oxidation by UV/sunlight, activated carbon adsorption, chemical oxidation etc., have been conventionally used as remediation technologies for decontamination of such sites, however, these techniques have several limitations and are not suitable for in-siru treatment of explosive contaminated sites (Burrows, W.D. 1982, Donaldo & Freeman, 1991, Hawthorne *et al.*, 2000). Further they also lead to recompartmentalization of pollutants from one stream into another resulting in magnifying the pollution load in the environment.

At such a crux, bioremediation technology, in particular phyto-remediation comes as a savior for in-situ treatment of explosives contaminated sites (Chen, 1993; Moffat, 1995; Schnoor *et al.*, 1999).

Phyto-remediation is a green technology, which involves in-situ treatment of contaminated soils, sediments, and water using vegetation. It is best applied at sites with shallow contamination of organic, nutrient, or meal pollutants that are amenable to one of five applications: Phyto-transformation, Rhizosphere Bioremediation, Phyto-stabilization, Phyto-extraction or Rhizofiltration.

Some plants are able to extract and concentrate particular elements from the environment, thereby offering a permanent means of remediation for variety of substance comprising explosive (RDX, HMX, TNT, DNT, TNB) (Nandkumar etc., 1995); heavy metals (Cd, Cr (VI), Pb, Co, Cu, Ni, Se, Zn) (10); radio nuclides (Cs, Sr, U) (12); petroleum hydrocarbons (BTEX); polychlorinated biphenyls (PCBs); chlorinated pesticides; organophosphate pesticides (e.g. parathion (13-14); nutrient (nitrate, ammonium, phosphate) and Surfactants. Although, Phyto-remediation may take longer than traditional approaches to reach cleanup goals, still as a rule, plants will survive higher concentrations of hazardous wastes than will most microorganisms used for bioremediation as it involves integrated mechanism. Therefore, owing to its simplicity and cost effectiveness, Phyto-remediation is bound to carve a niche in the time to come.

The study was carried out with the aim to investigate feasibility of Phytoremediation technology for the removal, of *Ocimum* from the soil under simulates and controlled environmental conditions using commonly available plants: Sacred basil (*Ocimum sanctum*), Black grass (*Ophiopogon* sps) and tomato (*Lycopersicon* sps.). Results of this study revealed phytoavailability of RDX and HMX and their possibility of removal from the soil by plants *Ophiopogon* sps (Black grass) is a versatile plant species that can with stand higher concentrations of RDX and HMX upto 301mg/kg, *Lycopersicon* sps. (Tomato) showed a high removal of almost 96% at a concentration of 25 mg/kg and significant removal rates at other concentrations of 37.5 and 50 mg/kg. *Ocimum* sps. is very sensiitive to the concentration of both RDX and HMX. Conjugate removal mechanism operates in the soil by involvement of both biotic and abiotic factors and is greener and ecofriendly option to tackel such wastes. Phytoremediation is an apt adoption at explosives contaminates sites after extensive geo-morphological study.

5.3 Phytoremediation of Water Pollutants

Water and plant samples were taken near the dumpsites, which is about 500 m away from the creek. Results of the water analysis showed that the dumpsite and Panlasian Creek were slightly polluted with considerable amount of phosphate. Results of the plant chemical analysis showed that kangkong (*Ipomea aquatic*) and Hydracharitaceae (*Ottelia alismoides* L.) were both efficient in phytoremediating Pb. Analysis of the plants further suggested that the concentrations of Pb in morning glory (*Ipomea violacea* L.) and hydracharitaceae (*Ottelia alismoides* L.) and hydracharitaceae (*Ottelia alismoides* L.) was about 210% more than the concentration of Pb in the water (Xia and Ma, 2005).

Xia and Ma (2005) had investigated the potential of water hyacinth (*Eichhornia crassipes)* in removing a phosphorus pesticide ethion. The disappearance rate constants of ethion in culture solutions were -0.01059. -0.00930, -0.00294 and -0.00201 for the non-sterile planted, sterile planted, non-sterile unplanted and sterile unplanted treatment, respectively. The accumulated ethion in live water hyacinth plant decreased by 55% - 91% in shoots and 75% - 81% in roots after the plant growing 1 week in ethion free culture solutions, suggesting that the plant uptake and phytodegradation might be the dominant process for ethion removal by the plant. Given the promising result of the study, water hyacinth could be utilized as an efficient, economical and ecological alternative to accelerate the removal and degradation of agro-industrial wastewater polluted with ethion.

Letachowicz *et al.,* conducted a study on the phytoremediation capacity on heavy metals accumulation in different organs of *Typhia latifolia* L. The concentrations of Cd, Pb, Cu, Ni, Mn, Zn and Fe were determined in different organs of *Typhia latifolia* from seven water bodies in the Nysa region in Poland. The *Typhia latifolia* species that can absorb heavy metals can be used as bio-indicator of pollutants is a macrohydrophyte and is widely present in the entire lowland and lower mountain sites. It is linked with nutritious water and organic or inorganic mineral bottom sediments.

5.4 Phytoremediation of Mining Sites

The focus of this study were on the accumulation of heavy metals in plants most commonly found in mine tailings of Victoria, Manlayan, Benguet, Philippines and identification of the different plant species within the area of the study. These plant species were assumed to be potential phytoremediation species (Wislocka, Krewezyk, Klink and Morrison, 2006) and also AM fungal association on coastal plants are considered to be vital in nature (Lakshman, 2000; Lakshman and Inchal, 2000).

The heavy metals extracted from the plants in the mine tailing were Cu, Cd, Pb and Zn. The fourteen plant species that were identified within the study were; *Eleusine indica* L.; *Amaranthus spinosus* L.; *Alternathera sessilis* L; *Portulaca oleracea* L.; *Fimbristylis meliacea* L. Vahl; *Mikania cordata* (Burm. F.) B.I. Robins; *Polygonum barbatum* L.; *Achyranthes aspera* L., *Blumea* sp., *Cyperus alternifolius* L.; *Crassocephalum crepidioides* (Benth.) S. Moore; *Cyperus compactus* Retz; *Desmodium* sp. *and Muntingia calabura* L. These plants absorbed certain metals at low and high levels. Among the plants species, *A. spinosus* was found to have almost all the metals extracted in large amount particularly Pb. The other plant species with high concentration of Pb were *A. sessilis, Desmodium* sp., *P. oleracea,* and *A. aspera. E. indica* has the highest concentration of Zn together with *M. cordata, C. compactus, F. meliacea* and *A. spinosus.* In contrast, Cd was found in trace amount in soil, but high in the following species: *C. crepidioides, P. oleracea, A. Sessilis,* and *C. alternifolius.* Nickel was found high only in *A. Sessilis* and *Blumea* sp. but trace amount in *Desmodium* sp. and *F. meliacea.* Also, high Cu concentrations were found in *A. spinosus* and *P. aleracea.*

In this study, the phytoremediation potential was dependent on population within species. The potential of the surveyed species mentioned for phytoremediation was remarkable and promising because of the presence of heavy metals suspected to have accumulated in the soil. Root system of these plants showed higher root to shoot ratios compared to other plants found in the area indicating high translocation of metals to the shoot was recorded in tropical herbs (Lakshman, 1992). These species also plays an important role in the phytostabilization of metals to reduce leaching and run off. Also, these may be transformed to less toxic forms. These typical plants have dense root systems which can be effective for phytostabilization and elimination of contaminants such as Pb, Cd, Zn, As, Cu and Ni in mine tailing sites.

A similar study conducted in Poland was worth including in this section. Wislocka *et al.,* (2006) have studied the bioaccumulation of heavy metals by selected plants from uranium mining dumps in the Sudety Mountains, Poland. They found out that the investigated plants from the uranium dumps in the Sudety Mountains grew on acidic soils with an unfavorable C/N ratio. However, the nutrients status as well as relatively high CEC and organic matter of the soil allowed the growth of spontaneous vegetation. Contamination by heavy metals (Pb, Zn, Cu, Cd and Ni), being associated with the mineral assemblage of the spoil material, was found to be significant within all dumps. All plants examined (*Salix caprea, Betula pendula and Rubus idaeus*) accumulated high amounts of heavy metals, but in general *R. idaeus* showed lower concentration of heavy metals (except Mn) in its leaves. However, Pb was accumulated to a similar degree in both trees and *R. idaeus*. Among all the heavy metals analyzed in the three species, Cd exhibited the greatest accumulation rate and the Cd accumulation ratio was several times higher for *S. caprea*, in comparison to the other two species. *B. pendula* and *R. idaeus* exhibited higher accumulation rates for Mn and *S. caprea*. However, the potential use of *R.idaeus* in monitoring metal concentration in the environment requires further investigation. The significant positive correlation between Pb in soil and leaves of the same tree suggest that *S. caprea* should be employed for monitoring Pb in the environment. The aggregation and sand-dune soil by AM fungi, its use in revegetation reported by (Lakshman, 1999).

5.5 Phytoremediation species in Coastal Water

Mangroves are higher plants, which are found mostly in the intertidal areas of tropical and subtropical shorelines and show remarkable tolerance to high amounts of salt and oxygen poor soil. The mechanisms of mangrove to keep the salt away from the cytoplasm of the cell were through the excretion of salt in their salt glands found in the leaves and roots and through storage of salts in the mature leaves, bark and wood (Kathiresan and Bingham, 2011). Mangroves developed unique body features in order to cope up with harsh environment. There are different types of roots, such as prop roots in *Rhizophora*, pencil-like pneumatophores in *Avicennia,* and cone-like pneumatophores in *Sonneratia* that have large lenticels to permit gas exchange. The leaves of mangroves have characteristics to surive from dessication and conserve water like the presence of thick epidermis, waxy cuticle, and presence of hypostomata. Mangrove ecosystem is exposed to different pollutants such as heavy metal, sewage wastes, pesticides and petroleum products. Heavy metal accumulation in the mangrove sediment can result in biological and ecological effects. Even though, mangrove trees may have the immunity against the toxic effects of the heavy metals, but the animals thriving in the ecosystem are vulnerable to the negative effects of heavy metals (Zheng, Chen and Lin, 2012).

Few studies were conducted about phytoremediation potential of mangroves and other wetland plant species. However, those researchers paved the way to explore more species of mangroves particularly the native species present in the area, for their feasibility to accumulate heavy metals. Zeng *et al.*, (2012), had studied the different metal concentrations of Cu, Ni, Cr, Zn, Pb, Cd, and Mn in *Rhizospora stylosa* at Yingluo Bay, China. The study showed less pollution due to relatively low concentration of metals especially Pb, Mn, Zn and Cb.

MacFarlane and Burchett (2011) have examined the cellular distribution of Cu, Pb and Zn in grey mangrove, *Avicenia marina* (Forsk) using scanning electron microscope X-ray microanalysis and atomic absorption spectroscopy. They reported that metals mostly accumulate the plants cell walls. Their study showed that certain parts of mangroves have the ability to control the entrance of heavy metals in other parts of the plants. The laboratory research of MacFarlane and Burchett contributed information on the accumulation, growth effect and toxicity of Cu, Pb and Zn in grey mangrove, *Avicennia marina* (Forsk.). Accumulation of the different metals occurred at varying concentrations in the roots and leaf tissue with AM fungal colonization (Lakshman, 1999). In the roots, Pb accumulated lesser than the other metals while high concentration of Zn was found in the leaf tissue. The effects of excessive Cu and Zn on young mangrove were reductions in seedling height, leaf number, total biomass and root growth.

Saenger and McConchie (2011), have evaluated the accumulation trend of Pb in the tissues, barks and woods, old and young leaves and fruits of different mangrove species. They discovered that Pb concentrated more in the bark than in other tissues of mangroves because of atmospheric Pb due to vehicle exhausts from nearby major roads.

Shete *et al.*, (2011), have reveals in their study entitled, "Bio-accumulation of Zn and Pb in *Avicennia marina* (Forsk.) from urban areas of Mumbai (Bombay), India", that the mangrove species can bioaccumulate and survive despite heavy metal contamination. Results showed that mangroves have greater uptake of heavy metals. Variations on the concentrations of Zn were found from the different plant parts while high accumulation of Pb was focused in the roots. They found out that Pb concentrations were present in the leaves and roots. Kamaruzzaman *et al.*, (2011) studied the cumulative partitioning of Pb and Cu in the *Rhizosphora apiculata* in the Setiu mangrove forest, Terengganu. Results showed increasing concentration of Cu and Pb from the leaf, bark, root and sediments.The study by Pahalawattaarachchi *et al.*, (2011) reported the absorption, accumulation and partitioning of eight different metals specifically Cu, Cd, Cr, Fe, Mg, Ni, Pb and Zn by mangrove species, *Rhizosphora mucronata* (Lam.) at Alibag, Maharashtra, India. They revealed that Cu, Mn and Fe showed limited mobility due to their accumulation in the roots while other metals (Cd, Zn, Ni and Pb) were concentrated in the aerial part of the plant. They concluded that *Rhizosphora mucronata* (Lam) at Alibag was more capable of phytostabilization rather than phytoextraction because of low uptake capacity of different metals.

Nazli and Hashim (2011) have revealed that *Sonneratia caseolaris* was a potential phytoremediation species for selected heavy metals in Malaysian mangrove ecosystem. The study assessed the concentrations of Cd, Cr, Cu, Pb and Zn in *Sonneratia caseolaris*. Results showed that both roots and leaves of *Sonneratia casealaris* accumulated and exceeded the general normal upper range of Cu and Pb in plants. In Iran, Parvesh *et al.*, (2011), have studied the bioavailability of different heavy metals (Ni, Cu, Cd, Pb and Zn) in the sediments of Sirik Azini creek. The outcome of their research revealed no heavy metal pollution was found in the area due to low geo-accumulation index of Pb in the sediment. They assessed that the concentration of heavy metals particularly Pb in the leaves were higher than the concentration of Pb in the sediment.

Qui *et al.*, (2011), have studied the different accumulation and partitioning of seven trace metals, namely, As, Cd, Cr, Cu, Hg, Pb and Zn, in mangroves and sediments from three estuarine wetlands of Hainan Island, China. They analyzed the sediment samples and found out that the heavy metals present in the area were still at relatively low levels. Further more, Pb analysis of mangroves showed that this metal was found mostly in the branches of the different mangroves. Zhang *et al.*, (2011) had investigated the physiological response of *Sonneratia apetala* (Buch) to the addition of wastewater nutrients and heavy metals (Pb, Cd, and Hg). They planted mangroves in four different treatments;

(1) Control, which has only salted water;

(2) Normal concentration of wastewater nutrients and heavy metals;

(3) Five times the normal treatment; and

(4) Ten times the normal treatment.

Results revealed that growth of mangrove increased with increasing levels of wastewater pollution. The study showed that mangroves were potential phytoremediator in wetland ecosystem.

Research by Nirmal *et al.*, (2011) entitled, "An assessment of the accumulation potential of Pb, Zn and Cd by *Avicennia marina* (Forssk) in Vamleshwar Mangroves, Gujarat, India", reported that sediments in the area are below critical soil concentration for heavy metals. *A. marina* possesses the capacity to uptake selected heavy metals, Pb, Zn and Cd, via its roots and storing them in their leaves without any sign of complications.

5.6 Phytoremediation of Soil Pollutants

Phytoremediation is a cleanup technology for metal contaminated soils, specifically Pb. In order for this type of remediation strategy to be successful, it is necessary to utilize metal accumulating plants to extract environmentally toxic metals from the soil, such as Pb, Ni, Cr, Cd and Zn. Certain plants have been identified not only to accumulate metals in the plant roots, but also to translocation the accumulated metals from the root to the leaf and to the shoot. While many plants performed this function, some plants, known as, "hyperaccumulators", can accumulate extremely high concentrations of metals in their shoots (0.1% to 3% of their dry weight) (Huang and Cunningham, 1996). The metal rich plant material can then be harvested and removed from the site without extensive excavation, disposal costs and loss of top soil that is associated with traditional remediation practices.

Bioremediation process would be extremely slow because the rate of biomediation is directly proportaional to growth rate while the total amount of bioremediation is correlated with a plant's total biomass. No plant has been discovered yet capable of meeting all the ideal criteria of an effective phytoremediator. These criteria are fast growing, deep and extensive roots, high biomass, easy to harvest and hyperaccumulators of a wide range of toxic metals. A Pb absorption study by Huang and Cunningham (1996) had cited corn as a perfect phytoremediator due to its large biomass, fast rate of growth and the existence of extensive genomic knowledge of this crop. Introduction of hyper accumulating genes as well as genetic information would better prepare these species to deal with diverse climatic conditions (Clemens, Palmgren and Kramer, 2002). The mobilization of metal contaminants, both in the soil and the plant, is another important factor influencing the success of phytoremediation. The amount of soluble Pb^{2+} in the soil appears to be a key factor to the enhancement of Pb^{2+} uptake by plants was studied by Alloway (1990).

Two main amendment techniques have been used to increase the bioavailability of Pb in soils and the mobility of Pb within plant tissue by lowering soil pH and adding synthetic chelates. Soil pH is a significant parameter in the uptake of metal contaminants because soil pH value is one of the principal soil factors controlling metal availability (Huang, J. W., Chen, J., Berti, W. R. and Cunningham, S. D. 1997). Maintaining a moderately acidic pH in the soil may be attained through the use of ammonium containing fertilizers or soil acidifiers. By this, Pb metal bioavailability and plant uptake can increase (Salt, Blaylock, Chet, Dushenkov, Ensley, Nanda and Ruskin, 1995; Cholpecka, Bacon, Wilson and Kay, 1996). In a study performed by Cholpecka *et al.*,(1996), have studied on metal contaminated soils in southwest Poland, reported that soil samples with pH of less of less than 5.6 contained relatively more metals in the exchangeable form than in soil samples with pH greater than 5.6. In addition, at lower pH, the Pb in soil has a greater potential to translocate from a plant's roots into its shoots. Synthetic chelates, such as ethylenediaminetetraacetic acid (EDTA), have been shown to aid in the accumulation of Pb^{2+} in the tissue. EDTA and other chelates have been used in soils and nutrient solutions to increase the solubility of metal cations and the translocation of Pb into shoots.

The physiological and biological mechanisms involved in Pb uptake of plants involving root to shoot transport of Pb may require some time to develop and become functional. Since plant species can differ significantly in Pb uptake and translocation, the success of using plants to extract Pb from contaminated soils required the following; The identification of Pb accumulating plants that can survive in the presence of contaminants; The measurement of the concentration of pollutant in the soil; and Knowledge of chemistry (availability or speciation) of the metal in the soil matrix.

6
Role of Rhizosphere and Microbes in Phytoremediation

Introduction

The rhizosphere extends approximately 1 mm around the root and is under the influence of the plant. Plants release exudates in the rhizosphere likely to serve as carbon source for microbes (Bowen and Rovira, 1991). As a result, high microbial build up of 1-4 orders of magnitude occur in the rhizosphere compared to bulk soils (Olson et al., 2003). Consequently, rhizosphere microbes can promote plant health by stimulating root growth via production of plant growth regulators, enhance mineral and water uptake. Rhizosphere remediation may be a passive process. Pollutants can be phytostabilized simply by erosion control and hydraulic control. There is also passive absorption of organic pollutants and inorganic cations to plant surface. Pollutant adsorbed to lignin cells is called liginification.

Microbes and plant activities also affect pollutant bio-availability. Some bacteria release biosulfactants (rhamnolipids) that make hydrophobic pollutants more water soluble (Volkering et al., 1998). Plant exudates may also contain lipophilic compounds that increase pollutant water solubility or enhance biosulfactant-producing bacterial populations examined by (Siciliano and Germida, 1998). Organic pollutant may be degraded in the rhizosphere by root released plant enzymes or through phytostimulation of microbial degradation. Organics such as PAHs and PCBs and other petroleum hydrocarbons have successfully been remediated in the rhizosphere by microbial activity reported by (Hutchinson et al., 2003, Olson et al., 2003). Plants stimulate the entire process by firstly, releasing carbon compounds to facilitate a higher microbial population around root zone. Secondly, secondary plant compounds released from the roots may specifically induce microbial genes involved in the degradation or act as co-metabolite to facilitate microbial degradation recorded by (Olson et al., 2003; Leigh et al., 2002). Also, roots of leguminous plants that host bacteria species with potential to convert atmospheric N_2 to inorganic N_2

in the soil can improve the C:N ratio of hydrocarbon contaminated soils, which ultimately enhance the process of rhizodegradation. Nwoko *et al.*, (2007) have reported sustained plant growth, leaf area and biomass production in *Phaseolus vulgaris* grown on spent engine oil contaminated soil.

6.1 Rhizobia and Azotobacter

Phytoremediation through *Rhizobia* and *Azotobacter* inoculated plants in Saline Soils

The introduction of genetically engineered *Rhizobia* and *Azotobacter* which are salt and drought tolerant suits the condition at Sinai Governorate, where temperature is high and saline water is used to irrigation. The engineered strains and be also efficient in nitrogen fixation to reduce the cost of fertilization, soil pollutants and improve the soil properties. Scientists constructed some strains tolerating 10-20 per cent NaCl. These strains were originally obtained by isolation of the responsible genes from *Bacillus* sp. strain tolerating 30 per cent NaCl and 55°C which was in turn being isolated from the soil of Aswan Governorate. The introduced genes resulted in obtaining efficient nitrogen fixing strains of *Rhizobia* and *Azotobacter*. The strains were introduced in the cultivated areas in Sinai with the new introduced crops. The results of the yield and the nitrogen contents in plants were significantly higher than those of the control areas, where these strains were not applied (El-Saidi and Ali, 1993). Application of salt tolerant *Rhizobia* isolated from wild legumes for the cultivation of groundnut in saline soil with pH 9.5 was done by Gangwane and Salve (1993). In three trials in field conditions, they found that both the dry matter and total nitrogen of groundnut plants were increased due to nodulation by all rhizobial isolates, which were efficient in N-fixation. There was also increase in the pod yield from 16.9 to 26.2 per cent due to the application of these *Rhizobia* to the seeds at the time of sowing.

6.2 Nickel – Resistant bacteria

Rhizosphere of Nickel hyper accumulating plants: a niche for Nickel - Resistant Bacteria

Plants growing in presence of phytotoxic concentrations of heavy metals have adopted several resistance strategies and acquired different mechanisms for growth. The most important one is the accumulation of extremely large amounts of heavy metals in their roots as well shoot system. Hyper accumulating plants having a high biomass production limit along with capability to concentrate 1-2% of metal in the above-ground portion are particularly interesting in phytoremediation technologies for treatment of metal-polluted soils, sediments and water bodies. The use of metal hyper accumulators in phytoextraction of metals has received much attention due to the possibility of decontaminating some of studied (Rawlings, 2002).

Several plant species endemic to metalliferrous soils have been identified as potent hyper accumulators, out of which 75% are able to concentrate nickel when growing in ultramafic soils. Ultramafic or serpentine soils are natural ecosystem developed over rocks rich in ferromangnesium minerals. These soils are derived through weathering and pedogenesis of serpentinite minerals and are enriched with nickel, cobalt and chromium besides iron and magnesium documented by (Brooks, 1987). The metal-stressed soil is poor in essential plant nutrients, nitrogen, phosphorus and potassium and supports endemic vegetation consisting of magnesium restrictors, calcium extractors, pathogens escapes and metal-resistant species or metallophytes/serpentinophytes (Kruckeberg, 1992). Ultramafic outcrops occurring around the world are associated with rare endemic serpentinophytes most of which are nickel-metallophytes, possessing the ability to accumulate over 1000 µg nickel/g of their foliar dry matter (hyperaccumulators). Vegetation of serpentine soil is sparse showing open stands of stunted and bushy shrubs and xeromorphism. Because of the occurrence of serpentinophytes, the number of endemic taxa are high in serpentine out crops and the ecosystem is interesting with respect to biodiversity and as a unique ecological island (Wenzel *et al.*, 2003).

Several studies have indicated that rhizosphere processes such as root architecture, effective root uptake systems and partial depletion of labile metals in the rhizosphere of hyper accumulators may play important role of root exudates in metal mobilization is still not clear. Recently it has been shown that rhizosphere microorganisms increase the availability of heavy metals for plant uptake (Amir and Pineau, 2003). Microbes are not only ubiquitous in metal-polluted environments (Basu *et al.*, 1997; Roane and Kellogg, 1996), but also in metal-percolated soils like the naturally occurring metalliferrous serpentine ecosystems. According to Hughes and Poole (1989) soil microorganisms are known to play key role in mobilization as well as immobilization of heavy metals. Microbial communities in the rhizosphere of metallophytes in heavy metal-contaminated sites have been reported to show efficient capability of detoxification of polluting substances by rendering them insoluble. However, in spite of the increasing knowledge regarding metal-microbe interactions, few studies have been attempted to characterize the significant role of nickel- resistant bacterial population in rhizosphere of nickel-hyperaccumulators in serpentine soils (Amir and Pineau, 2003; Idris *et al.*, 2004).

Schlegel *et al.*, (1991), have hypothesized that the leaves of this tree drives a "nickel cycle" which involved the transportation of nickel from deep soil and rock layers to the leaves, where accumulated metal is transiently stored in the vacuoles. After the leaves fall the nickel is liberated from the decaying leaves and is percolated by the rain water through the top soil layers, resulting in contents of about 10 mg nickel per gram of soil. This nickel cycle has led to the evolution of a higher percentage of nickel- resistant bacteria in soil under the canopy of nickel-hyper accumulating trees.

6.3 Arbuscular mycorrhizae and saline conditions

Phytoremediation through Arbuscular Mycorrhizal Fungi (AMF) under saline conditions

Arbuscular mycorrhizae refer to those mycorrhizal associations formed with fungi in the family *Endogonaceae*. Vesicular-arbuscular mycorrhizal fungi are aseptate phycomycetes. The fungi cannot be cultured on synthetic media, so they must be multiplied in association with a host plant. These fungi produce structures known as vesicles and arbuscules, in addition to hyphae and spores. Arbuscules are intracellular, haustoria – like structures that develolp by repeated, dichotomous branching of hyphae. Arbuscules can be seen with a light microscope and are short-lived; varying from one to three weeks. They can form very soon after the roots become colonized. Vesicles are sac – like, usually terminal swelling at the tip of hyphae.

The role of VAM (Vesicular-arbuscular mycorrhiza) in nutrition of agricultural and horticultural plants has received much attention, with particular emphasis on phosphorous nutrition. It is likely that effects of mycorrhizae in increasing nutrient uptake will be most marked for nutrients, which move to roots principally by diffusion and for host plants with coarse roots and few hairs. The reason that most attention has been given to the role of VAM (Vesicular-arbuscular mycorrhiza) a non-renewable resource, therefore efforts should be directed towards more efficient utilization of this resource. The VAM (Vesicular-arbuscular mycorrhizal) fungi play a significant role in this respect, particularly under different stress condition.

Interaction between host plant and fungus are complex and include symbiotic relationships. The host plant supplies carbohydrates to the mycorrhizal fungi, while benefitting from the fungal relationship primarily through enhanced uptake of immobile, inorganic nutrients. The mycorrhizal relationship with the host plant also can include increased resistance of the host to diseases, drought and salinity. AM fungi also are reported to enhance N_2 fixation in legumes (Hayman, 1986).

Future research is required for the identification of efficient AM endophyte species. Studies on symbiotic effectiveness, competitiveness against indigenous AM fungi, tolerance to environmental extremes such as drought and salinity, tolerance to mineral deficiencies, and toxicities, as well as tolerance to a broad range of pesticides are also needed.

Several reports dealing with the possible role of AM fungi in reducing the stress effect on the host are available (Safit *et al.*, 1972; Hirrel and Gerdmann, 1980 and Ojala *et al.*, 1982). However, the occurrence of AM in saline soils and their influence on plant growth and nutrient uptake has received little attention by researchers. Halophytic and non-halophytic plants are often mycorrhizal in calcareous (Read *et al.*, 1976) and salt-affected soils (Khan, 1974 and Malibari *et al.*, 1990). Plants inoculated with AM fungi had more dry weight yields P concentration in shoot and root tissues and total P uptake than non-mycorrhizal plants growing in salt-affected soil

6.4 Mycorrhizae and heavy metals

Phytoremediation through mycorrhiza inoculated Plants to remove heavy metals

Heavy metal contamination caused by either natural process or by human activities is one of the most serious environmental problems (Reedy and Prasad, 1990). Because plants function as the principal entry points of heavy metals into the food chain leading to animal and man (Rauser, 1990) is very important to analyse the distribution of different heavy metals in plants. Since most plants are mycorrhizal under ecological conditions, it seems appropriate to review interactions of mycorrhizal fungi with heavy metals. Since mycorrhizal fungi play an important role in the establishment of plants under adverse soil conditions, the study of interactions of mycorrhizal fungi with heavy metals is important so as to explore the possibility of reclaiming heavy metal contaminated lands. Ecto-and endo-mycorrhizal symbiosis plays a crucial role in connecting plant roots with the soil (Harley, 1978).

The higher absorption of nutrients by mycorrhizal plants is usually attributed to extended extrametrical mycelium that provides exploration of soil volume (Galli *et al.*, 1994). It was also shown that the uptake of heavy metal micronutrients such as Cu, Zn and Co present in soil solution at low concentrations was enhanced by mycorrhizal infecti on examined by (Harley and Smith 1983; Lambert *et al.*, 1979; Li *et al.*, 1991). However, when heavy metal micronutrients are present at high concentration may be disadvantageous for mycorrhizal plants. This has been demonstrated with the bunchgrass, *Erharta calycina* inoculated with AM fungus, *Glomus fasciculatum* at low substrate pH observed by (Killham and Fireston, 1983). On the other hand, association with mycorrhizal fungi can protect plants from heavy metals. While a wide range of responses have been reported, in most cases, some degree of host amelioration of metal toxicity by mycorrhizal fungi has been demonstrated (Colpaert and VanAssche 1992 a, b; Hartley *et al.*, 1997; Marschmer *et al.*, 1996). Bradely *et al.*, 1981) could show an improved heavy metal resistance of ericoid mycorrhizal *Calleuna vulgaris*, in comparison with non-mycorrhizal controls. Gildon and Tinker (1981), found a heavy metal tolerant strain of AM fungus *Glomus mosseae*, which was collected from a heavily Zn and Cd contaminated site. Rufyikiri *et al.*, (2000) reported a decrease in aluminium content of roots and shoots of arbuscular mycorrhizal plants as compared to non-mycorrhizal banana plants.

Brown and Wilkins (1985 a, b), have demonstrated that birch seedlings, grown from seeds collected from trees at a zinc-polluted site, showed a greater metal tolerance than specimens collected from non-polluted sites. However, these authors could not find correlation between the tolerance of their mycorrhizal fungi, *Amanita muscaria* and *Paxillus involutus* and the toxicity of the soil of origin (Brown and Wilkins 1985 a, b). Effect of external concentration of Zn on its growth and uptake by *Betula* spp. And the mechanism of EM fungal amelioration of Zn toxicity in Betula spp. Investigated by Denny and Wilkins (1987 a, b). Ectomycorrhizal Pinus banksiana and *Picea gluca* infected with *Suillus luteus* have been found to be more tolerant to heavy metals present in soil (Dixon and Buschena, 1988).

Lo-Buglio and Wilcox (1988), have reported that the survival and growth of EM and ectoendomycorrhizal red pine seedlings transplanted from nursery onto iron tailings was better than the controls. Aluminium has also been shown to influence mycorrhiza formation. *Pinus caribaea* var. hondurensis was inoculated with seven species of EM fungi in the presence of six levels of Al toxicity at all the concentrations whereas. *P. tinctorius* and *Suillus* sp. were depressed at lower but stimulated at higher Al levels, and the inverse was true for *Rhizopogon reaii* and *M. cylindrosporum*. The EM fungi also appeared to ameliorate Al damage to plant roots even in treatments where no formation of mycorrhiza (Kasuya *et al.*, 1990). The seedings mycorrhizal with *P. tinctorius* exhibited unaltered growth and physiological function (Cumming and Weinstein, 1990). Hartley *et al.*, (1998), have reported that Scots pine was considerably more insensitive than the fungus to Cd-Zn in vermiculite culture media, yet the fungus still benefited the plant at low infection levels. Six strains of EM fungi including 'pioneer' and 'late stage' fungi as well as one strain isolated from Zn-polluted site were comparable for their ability to increase Zn tolerance in *Pinus sylvestris* seedlings. The mycobionts varied considerably in their protection of the autobiont against zinc.

Toxicity and Zn accumulation in the seedlings - Mycorrhizal fungi act as a safety net that can immobilize large amounts of Zn, thus preventing transport to the host plant (Colpaert and VanAssche, 1992 a). Nine EM fungal strains collected from a Zn and Cd polluted soil or from an unpolluted area was comparable for their protective effect on ectomycorrhizal *Pinus sylvestris* against toxicity at a sublethal Cd concentration of 445µM. A protective effect in the host was observed with all mycobionts, since Cd uptake was highest in the non-mycorrhizal seedlings. Mycorrhizal fungi have also been reported to reduce manganese toxicity to *Vaccinium macrocarpon* in perlite culture containing nutrient solution amended with Mn at 0, 250, 500 and 1000g/mL (Hashem, 1995). The Zn and Cd tolerance of *Suilus luteus* isolates from the polluted habitat were significantly higher than the tolerance of isolates from non-polluted site was considerably larger than that observed at the polluted sites. The influence of EM fungi on the uptake toxicity of metals has been extensively reviewed by Wilkins (1991) and Hartely *et al.*, (1997).

Interspecific variation exists in the ability of EM fungi to confer a reduction in metal sensitivity to host plant, and this depends on mycorrhizal fungal, species, heavy metals ad the experimental conditions used. Hartley *et al.*, (1997), Berry (1982) and Marx (1975) reported that the inoculation of several species of *Pinus* with *P. tinctonus* resulted in better growth on strip mine spoil than did the naturally occurring *Thelephora* spp. *Scleroderma flavidum* was found to be the only fugus, out of four tested, which was able to increase Ni tolerance in *Betula papyrifera* and neither of the fungi affected Cu-toxicity (Jones and Hutchinson, 1986). Soil pH plays an important role in metal toxicity of plants. Low soil pH combined with high aluminium concentration has been suggested to get a major cause of forest decline in the North Eastern United States and Central Europe (Hooker and Shoomaker, 1985; Huttermanna, 1985). Entry *et al.*, (1987), have confirmed the above hypothesis by inoculating *Abies balsamea* with *C. geophillum, H.crustuliniforme* and *L. laccata*, and growing them in pots amended with 50 mg Al/g soil and 100 mg A1/g soil with pH adjusted to 3, 4 and 5. As the solution acidity and Al concentration increased, a decrease in EM formation, root weight, shoot weight, total weight, and root shoot ratio was observed (Entry *et. al.*, 1987). On the contrary in a similar study, Godbold *et al.*, (1995) have reported a reduction in the effect of Al on root growth and Al concentration in roots of *Picea abies* at low. Toxicity symptoms appearing on shoots are likely to be a consequence of root growth inhibition rather than a direct effect of toxic metals on shoot.

6.5 AM Fungi and Forest Nursery Seedlings

Importance of AM Fungi in the Improvement of Forest Nursery Seedlings

Trappe (1987), have reported so felicitously put it, `Most woody plants require mycorrhizae to survive, and herbaceous plants need them to thrive', we envisage increasing exploitation of both AM fungi. Man, after all, requires food to survive, and lumber to thrive.

The world's endo-mycorrhizal rainforests are rapidly being destroyed by extraction of timber, slash and burn agriculture, and other processes, all, unhappily, anthropogenic. Since these forests do not regenerate easily or quickly, it is very likely that in many areas they will be replaced by plantations of exotic conifers which often grow much faster than indigenous trees. Such plantations may also be initiated on many other currently treeless areas. And existing forests need to be re-established after harvesting. In every case, out biotechnological investments in arbuscular ectomycorrizal fungi will be well repaid by the ease and speed with which the trees become established, and by the grains in growth of the maturing forests. Tree nurseries and field crops are often heavily fertilized. This effectively discourages many mycorrhizal fungi, which are successfully weaned forest nursery men and farmers away from, the chemicals that have served them so well (Varma, 1995b). In the long run, increasing energy costs may provide the required impetus.

Horticulticulturists commonly use their own mixture of sand, peat and wood shavings as a propagating medium for cuttings, often adding lime and fertilizer. Needless to say, the mix does not contain AM inoculums. The addition of AM fungi would confer all the usual benefits upon these plants when they were out planted and would have the added advantage of reducing the amount of fertilizer that must be added to the mixture. Lakshman (2001), have suggested that AM fungi could be added to any substrate used in the green house to produce stock for transplanting to the field. At present, tomatoes raised in green houses have their root systems plunged in fertilizer before out planting. This step may eventually be replaced by plunging in AM inoculums.

Starter cultures of AM fungi have been available in the past from universities and from industry, but the difficulties associated with maintaining the purity of the inoculums have been major obstacles to their commercialization. Although pure inoculums of a single species is often used in laboratory experiments, this might not be not indeed be desirable, for field applications. A special mixture of selected AM fungi was recently placed on the market by Biofertec Inc. in Canada. The producers claimed that this mixed inoculums can make most species of garden plants, fruit trees and experience of this product, and we can neither confirm nor refute the claims made for it. It is certainly possible that amixture of several fungi might constitute a more generally applicable inoculums by increasing the chance that at least one of the potential mycobionts would be well suited to the particular host-soil-climate combination (Peterson Farquar, 1994)

This treatment would have damaged the hyphae, but not the spores or infected roots. Hayman (1987), have suggested that direct drilling and minimum tillage might reduce disruption of the mycelial network and favour established populations of AM fungi. It seems to us that this knowledge can be turned to good account in another way. When we need to introduce highly efficient strains, through cultivation should actually help to become established be cutting down competition from indigenous organisms.

Another way of doing this, or of eliminating known pathogens, is to fumigate the soil with methyl 1 bromide, indeed, in California the law requires that be done when plants destined for sale are to be produced. This treatment is extremely effective in eradicating soil microorganisms, and once the soil has been detoxified for the recommended time, introduced mycorrhizal fungi can proliferate without competition. Soil fumigation does, however, require equipment, time and money and may not always be necessary.

For many years, substances with specific toxicity toward fungi have been deliberately introduced to the plant or to the soil in order to control pathogens. Most early fungicides were toxic to a wide range of fungi, and also too many non-target organisms, not excluding the very plants they designed to protect. Members of the second generation of fungicides are far less phototoxic, and are also systemic. They enter the plant but remain apoplastic, moving only upward in the transpiration stream. Third generation fungicides are not only systemic, but symplastic, moving upward in the transpiration stream and downward in the phloem. The early fungicides posed a threat to mycorrhizal fungi only when deliberately applied to the seed or the soil, or when washed down into the soil. The most recent fungicides will reach the interior of the roots no matter where the application is made. Obviously, there is concern that these newer fungicides, in addition to controlling the target pathogen, may deter damage or displace the desirable mycorrhizal fungi present in or near the roots. Fortunately, some of the most recently introduced fungicides are highly selective in their action, and have negligible toxicity to many non-target organisms. In fact recently been demonstrated that foliar applications of a selectively oomycete fungicides, Aliette, actually stimulate the colonization of the roots of onion by a species of *Glomus*. The possible impact on indigenous or introduced mycorrhizal fungi of the wide range of biocides currently used in agriculture and forestry has not usually been considered in the past, but we believe it should be a subject for ongoing research, and a concern in the choice of strategies for the control of pest and pathogens.

The possible role of mycorrhizae in the management of marginally productive lands, such as arid and semi-arid rangelands where herd of animals graze, has been considered. Overgrazing is deleterious to the mycorrhizal system. Perhaps because the animals selectively eat forage plants that are mycorrhizal annuals, leaving behind unpalatable, non-mycorrhizal perennials. Non-disruptive management practices, such as the introduction of efficient mycorrhizal fungi, removal of non-mycorrhizal perennials and planting of mycorrhizal perennials, might result in higher productivity from such marginal lands. AM fungal inoculation with different concentration of domestic sewage water improving plant growth, so indirectly removing waste reported by (Kolkar and Lakshman, 2010).

Surveying the AM fungi of citrus orchards in California and Florida, found that plants of different ages tended to have different AM symbionts. They concluded that the age of the plant plays a role in determining which AM fungus will be the dominant mycorrhizal partner. It seems quite likely that under natural conditions, over the long term, a succession of AM fungi will associate with a single perennial plant. Since the source of inoculum for the later stages of this succession will in all probability be the community of AM fungi in the surrounding soil, programs aiming at introducing efficient AM fungi to perennials should concentrate on those best adapted to the earliest stages of growth; those which will bring about quick and extensive infection, and increase the probability of the young plant's survival. The spores of AM fungal germinate and if the germ tube or hyphae encounters a receptive root or root hair it forms an appresorium (Hall, 1977) the hyphal branch from the appresorium penetrates the epidermis in the zone of differentiation and elongation of feeder roots. The root tips are not normally infected.

The hyphae grow between the cortical cells and never enter the meristematic cells and the endodermal cells. Specialized branches enter individual cortical cells and form finely arborescent structures called "arbuscules" which are the sites of exchange between the fungus and the plant individual arbuscular function for 4-15 days. The progress of AMF colonization in a single root system commonly follows a sigmoid curve. The mathematical modeling of progressive measurement of infection has highlighted the need for more detailed experimental study of biology of individual infection. The development of fungus inside the root constitute intramatrical phase and the proliferation outside in the soil is extrametrical phase. The latter phase enables the plant to exploit greater volumes of soil available to a non mycorrhizal plant. The fungal efficiency depends upon the extent of extrametrical hyphae which in turn is based on the availability of the substrate in root. The role of volatile substances emitted by host roofs and gibberellins and cytokinin like substances secreted by the spore of fungus (*Glomus mossae*). Barea and Azocon Aquilar (1982), has recognized the colonization under natural conditions deserves further investigation.

6.6 Role of AM Fungi in waste land development

Large areas of land in the country called wastelands are degraded and lying unutilized due to various constraints. Part of it called culturable, has the potential for the development of vegetation cover. On the other hand, part of it called unculturable, has no potential for the development of vegetation cover. In addition to these natural wastelands, there are large tracts of wastelands created due to human activities such as mining, deforestation etc. Such wastelands are also culturable and vegetation cover can be developed on them after removing the constraints with some efforts. In view of the increasing shortage of plant re-sources due to population explosion it has become imperative that all the wastelands are put to use by developing vegetation cover. A variety of plantation programmes are being executed in the country utilizing known scientific inputs. Attempts are being made world over to utilize AM technology in plantation programmes. Indian mycologists have also made important contributions in this direction. Following description gives an account of their contributions:

Thatoi *et al.*, (1993), have investigated the comparative effect of AM inoculations with *Glomus mosseae* or *G. fasciculatum* on growth, nodulation and *Rhizobium* population of subabul (*Leucaena leucocephala*) grown in iron mine waste soil (from Orissa). Seedlings were grown in poly bags in 3 media: garden soil: mine soil: or mine waste soil mixed in equal proportions with sand. In general, inoculation with *G. mosseae* improved growth, nodulation and the rhizosphere rhizobial population, while inoculation with *G. fasciculatum* had negative effects. However, in pure mine soil inoculation with *G. mosseae* reduced nodulation and rhizophere rhizobial population.

Bisen *et al.*, (1996), has studied the AM colonization in tree species planted in Cu, Al and Coal mines of Madhya Pradesh with special reference to *Glomus mosseae*. Soils from the mine were acidic (pH 5.3-6.8) and low in available N, P and K, while soil from a non-mining area (Bhopa) was alkaline (pH 8.2). Most of the planted tree spp. showed colonization by AM fungi, with *G. mosseae* predominating. Colonization was highest on *Dalbergia sissoo* planted on a copper mine site (73%) and *Eucalyptus* spp. In a coal mine area (68%). Mining activities increased the abundance of AM fungi by lowering the pH.

Mehrotra (1996), has examined the potential use of revegetated coal mine spoil as source of Arbuscular mycorrhizal inoculums for nursery inoculations. Rhizosphere soils of five tree species were used as sources of mycorrhizal inoculums. Soils contained seven spore-forming species of AM fungi. The substrate used in the pot experiment was a mixture of unsterilized coal mine spoil without any mycorrhizal propagule and autoclaved sandy loam soil. *Cassia siamea* and *Derris indica* were used as the test plants. Measurements were made of shoot and root biomass, p uptake, per cent mycorrhizal infection and spore population of AM fungi. Growth measured as shoot and root dry weight was significantly higher in seedlings inoculated with soil inoculums from *Dalbergia sissoo*, *C. siamea*, *D. indica* and *A. indica*. *Acaulospora scrobiculata* was found to be the best fungus in terms of root colonization ability and effectiveness to promote P uptake and growth in plants. The results of the study justify the use of revegetated coal mine spoil as an effective and economical source of endomycorrhizal inoculum for inoculating nursery seedlings.

Chandra *et al.*, (1997), made a comparative study of Arbuscular-mycorrhizal fungal root infection and spore density in *Grevilia pteridifolia* in undisturbed and disturbed areas. Comparative studies of AM in two localities in undisturbed (nursery-site plantation) and disturbed (coal mine overburden) plantation of *G. pteridifolia* were undertaken. Mining operations considerably disturbed the AM population in coal mine overburden spoil. Highest percentage of root infection (19.36%) with spore density 412 spores/100 g soils was recorded in 1991 plantation of undisturbed soil in TFRI, Jabalpur. However, the infection of AM was 13.88% in 1991 plantation in coal mine overburden (with spore density 397 spores/100g soil). The infection of AMF in roots of *G. pteridifolia* and spores in soil increased with the age of the plants. The *Glomus* spp. was dominant in undisturbed plantations while *Acaulospora* spp. was more in mine site. The population of *Scutellopora* and *Sclerocystis* spp. was scanty in coal mine spoils. There was no definite trend in occurrence of AM fungi in undisturbed and disturbed plantations. This study confirmed that *G.*

pteridifolia is Arbuscular-mycorrhizal and it derives benefit from obligate symbiotic system thereby showing better performance in disturbed sites.

Rao *et al.*, (1999a) have, studied the role of AM fungi in rehabilitation of mine spoils. Surface mining is one of the important activities in arid region of Western Rajasthan which destroy vegetation. *Glomus fasciculatum* isolated from these areas was found to be the most effective than other strains. Out of thirty one plant species occurring naturally on these spoils, 10 species were selected based on their economic value to establish in mined areas after inoculation with the isolated *Glomus fasciculatum* species @ 10 spores/g soil. Per cent root infection increased by 2-9 folds in inoculated plants. Shoot biomass, N, P, K, Zn and Cu concentrations were significantly improved than normal in all the cases. The results demonstrated the possibilities of employing identified AM-fungi for rehabilitating mine spoils. The role of AM fungi and mine spoil consortium; a microbial approach on prosomillet and Fox tail millet by using mined spoil for increasing biomass yield was studied by (Channabasava and Lakshman, 2012; Lakshman and Channabasava 2013).

Thatoi *et al.*, (1999), had conducted experiment for the selection of efficient AM fungi for improvement of growth and nodulation of *Acacia nilotica* in chromite mine waste soil. Comparative assessment on the effect of 4 AMF species viz, *G. fasciculatum, Gigaspora gilmorei* and *Gigaspora margarita* on growth, nodulation and total N content of *Acacia nilotica* was done with a view to select out most efficient AM fungi for mine area soil. Results indicated a general increase in shoot height, plant dry weight and total N content of the plant due to all types of AM fungal inoculation but varied in their degree of stimulating capacity. Among four AMF species tried, *G. mosseae* showed highest stimulatory effect in terms of shoot height, plant dry weight and nodule numbers while *Gigaspora margarita* treatment showed least impact on improving biomass whereas *G. gilmorei* on nodulation of the test plant. Considering the overall effect on growth, nodulation and total nitrogen content of the plant. *G. mosseae* was found to be best species for inoculating *A. nilotica* in chromite soil. The effect of AM fungi and saline water with phosphorus was studied on Elusine coracana by (Patil and Lakshman, 2003).

Chandra *et al.*, (2000b) had made an effort to manage the wastelands via petroleum plantations and AM fungal technology. The study with soils of five different wastelands and four promising petrocrops clearly showed the promise of AMF technology in improving the quality and establishment of petrocrops in wastelands and problematic soils. Out of 26 petrocrops surveyed viz., *Calotropis procera, Catharanthus roseus, Euphorbia tirucalli, pedilanthus tithymaloides* var. *aureus, P. tithymaloides* var *cuculatus. P. tithymaloides* var. *variegates, Pergularia daemia and Sonchus asper,* showed highest potential to develop mycorrhization in their roots. The conclusions drawn from the study may prove appropriate guidelines for future work. The petrocrops have potentiality to form mycorrizal association with AM fungi under natural conditions. The magnitude of mycorrhizal status varies with the petrocrops as well as soil and environment related factors. Introduction of AM inoculants improves not only the mycorrhizal status of the petrocrops but also their biomass and biocrude content. The magnitude of improvement varies with the petrocrops and AMF incoculants. Appropriate AM fungi inoculants have the potentiality to help the petrocrops to overcome various types of constraints in wastelands and give better performance with reference to biomass and biocrude content. Cultivation of selected petrocrops may be undertaken in wastelands with AMF technology.

Sengupta and Chaudhuri (1995), have studied the effect of dual inoculation *Rhizobium* and mycorrhiza on growth response of *Sesbania grandiflora* L. in coastal saline and sand dune soil. They tested the response of single and dual inculations with *Glomus faciculatum* and salinity tolerant *Rhizobium* strain for *Senbania grandiflora* growing in pots in coastal saline sand due soil collected from the Sundarbans of West Bengal. After 120 days, both single and dual inculations gave significantly higher root and shoot dry matter yields over control, the non-incoculated plants. Growth difference between single inoculation of either *Rhizobium* or mycorrhizas was not significant but dual inoculations of both gave singificatly higher yields than single inoculations of either.

Madan *et al.*, (1995), have studied the effect of vesicular-arbuscular mycorrhizal fungi on different plant species in highly alkaline saline soils. Mycorrhizal inoculum (250g), consisting of 200 spores/100g soil besides fungal matrices, was inoculated into three and half month old seedlings of *Ailanthus excelsa*, *Pongamia glabra*, and *Cassia siamea*. No treatment was given to control. A standard amount of farmyard manure (10 tons/ha) was applied in all the cases. After four months, five plants from each plot were removed at random. Individual plants were taken out with care and their roots were washed gently under a fine jet of water, without damaging the root system. At the end of experiment, the soil was analyzed for AMF spore population. The data on the effect of AM fungi on the mortality and height of plants, the weights of stem and roots, and the percentage of AM colonization of roots showed that mycorrhizal inoculation resulted in 100% survival in all the three types of plants while in non-inoculated plants 10, 20, and 12% mortality was observed in the case of *A. excelsa*, *P. glabra*, and *C. siamea* respectively. Mycorrhiza-inoculated plants attained better shoot heights and shoot and root weights as compared to non-inoculated ones. The shoot dry weight of *A. excels*, *P. glabra* and *C.siamea* was enhanced by 28.4, 15.6 and 26.9%, respectively. A similar increase in root weight was observed in plants inoculated with AM fungi. The interaction studies revealed that the infection in roots was maximum in *A. excelsa* followed by *P. glabra* and *C. siamea*.

Sharma *et al.*, (1995a), have performed a study to identify efficient species/ strains of mycorhiza in enhancing survival, establishment and growth of *Acacia auricularis*, *Casurina equisetifloia* and *Pterocarpus marsupium* seedlings growing in degraded acid soils of Kerala. In all, they used 36 AMF cultures, 18 monospore cultures collected from the rhizosphere soils of the above trees and 18 obtained from other sources such as ICRISAT, BAIF, TNAU, USA, TERI and JNU. Out of these only 26 cultures were selected for glasshouse studies. High colonization of roots of *Acacia*, *Casuarina* and *Pterocarpus* by AM fungi indicated that these species are mycorrhiza-dependent. However, for each tree species, the efficient AM fungal species were different. Increase in growth, shoots and root dry weight and P content of inoculated plants was recorded which showed that the AM fungi may help the plants in their successful establishment in degraded soil. The effect of AM fungi on forest nursery seedlings was documented by (Waddar and Lakshman, 2010).

Vijaya *et al.*, (1996) have studied the sodic soil tolerance of teak (*Tectona grandis*) seedlings produced after gibberellic acid seed treatment. They planted seedlings in soil of pH 9.6 in pots, some of the pots and 5 g of the inoculums of *Glomus macrocarpus* mixed in the soil and seedlings survival, growth and leaf chlorophyll were recorded after 3 months. The mycorrhiza treated seedlings showed more tolerance to the sodic soil, better survival and exhibited significantly greater growth and chlorophyll content than controls.

Sharma *et al.*, (1996a), have searched out the possibility of using mycorrhizal and nitrogen fixing symbionts in reforestation of degraded acid soils of Kerala, Peechi. Several monosporal AMF cultures raised from spores isolated from the rhizosphere soils under *Acacia auriculiformis*, *Casurina equisetifolia*, and *Pterocarpus marsupium* and other AMF cultures obtained from various sources were screened under glasshouse conditions to select the most efficient ones for enhancing the growth of each of the three test tree species. However, response varied with different isolates of AM fungi. A significant increase in total P content was also observed in the inoculated plants. Based on overall performance and a cluster analysis of the data, the most efficient AM selected for the field studies were *Glomus caledonium* and *G. mosseae* for *Acacia*, *G. fasciculatum* for *Casurina* and *G. intraradices* and *G. mosseae* for *Pterocarps*. Forty cultures of *Rhizobium* and *Frankia* isolated from the nodules of *A. auriculiformis*, *P. marsupium* and *C. equisetifolia* were screened for efficiency in increasing growth and nodulating the 3 test tree species in the glasshouse. The two best strains were selected for each species.

In glasshouse trials the effect of fertilizers on seedling growth with and without symbionts, were also observed between various treatments (AM Fungi, N fixing symbionts and fertilizer) which showed significant difference in height and diameter of field planted 1 year old *A. auriculiformis*. All the treatments responded favorably to nitrogen and / or phosphours fertilizer application. The percentage increase in growth with fertilizer application was more in symbiont-inoculated plants than non-inoculated controls. Maximum percentage increase in height was 57.4 (*G. caledonium*+fertilizer) and 56.2 (*G. caledonium* + *Rhizobium* + fertilizer) whereas for diameter growth the maximum increase was 84.5% (*G. caledonium* + *Rhizobium*+ fertilizer) followed by 70% (*G. caledonium* + fertilizer). In control treatments the increase in height and diameter growth was 11.5 and 41.4 per cent respectively. Similarly, the effect of AM fungi and salinity on two legumes was recorded by (Romana and Lakshman, 2011).

Mishra (1994), has explored the possibility of using AM fungi in reclamation of wastelands of Eastern ghat regions of Orissa. Among the tree and crop legumes tested, *Sesbania grandis*, *Acacia grandis*, *A. nilotica* and *Leucaena leucocephala*, local crop legume *Cajanus cajan* cv. *Kandule* and *Vigna unguiculata* sub species *cylindrica* gave better performance when pre-inoculated with AM fungi.

Mishra *et al.*, (1995a), have studied the mycorrhizal development and plant growth in amended fly ash of thermal power plants. The population status of AM fungi in fly ash rehabilitated by planting with 6 multipurpose tree species (*Acacia auriculiformis, A. nilotica, Cassia siamea, Eucalyptus* hybrid (*E. tereticornis*), *Peltophorum ferrugineum* and *Pongamia pinnata* established in pits filled with soil (transported in or *in situ* and various organic amendments) at Chachai, Shahdol District, Madhya Pradesh, was compared with that under natural vegetation and in barren flyash. A pot culture experiment was also performed using fly ash/ soil/compost (1:1:1) as growing medium and 20 nitrogen fixing and non-nitrogen fixing tree species. Varying number of mycorrhizal spores were found in the 3 site types (none in barren fly ash, 50/100g soil under natural annual vegetation and up to 57/100g soil in the rehabilitated areas treated with transported soil and 52/100g soil in those with *in situ* soil) and in the pot experiment. There was a significant and positive correlation between percentage mycorrhizal infection and biomass production by the trees in the pot experiment. Mycorrhizal colonization of the trees in the rehabilitated area was 30-80% in transported soil, and 10-16% *in situ* soil. It has been suggested that selection and introduction of the mycorrhizal strains identified may assist future afforestation programmes, and help in the establishment of indigenous plants in rehabilitating areas.

Dixon *et al.*, (1997), have studied the Arbuscular mycorrhizal symbiosis in relation to forestation in arid land. Arbuscular mycorrhizal fungal inoculation of *Prosopis juliflora* seedlings using *Glomus macrocarpum* in the nursery of the University of Delhi significantly increased the juvenile growth response over 3 months. Three-month old-nursery-raised-mycorrhizal seedlings gave enhanced establishment (survival and growth) 3 months after out planting (i.e. at 6 months old) in 30 30 30 cm pits in nutrient deficient alkaline soil in Asala wildlife sanctuary (about 45 km from Delhi), a semiarid area with rhizosphere soil temperature of 52°C. The AM symbiont also increased drought tolerance of *Prosopis* plants in terms of maintaining and enhancing growth under water stress conditions. The results indicated the potential of Arbuscular mycorrhizae to partly reduce/replace the fertilizer requirements of trees in degraded, semiarid sites, where they can be described as acting as biofertilizers.

Kothari *et al.*, (1997), have studied the mycorrhizal biodiversity of petro effluent-irrigated fields. Petro effluent from the Indian Petro Chemicals Corporation (IPCL) has been tested for irrigation at the Ecoform at IPCL, Baroda, Gujarat, for recycling millions of tones of water. The rhizosphere of 6 crops raised in these fields was evaluated for mycorrhizal biodiversity combined with the mycorrhizal status of these crops. All the crops investigated showed efficient mycorrhization with Arbuscular mycorrhiza. The rhizosphere was rich in AMF species. *Acaulospora appendicula, A. birelicola, A. foveata, A. sporocarpia, Glomus aggregatum, G. deserticola, G. heterosporum, G. microcarpum* and 2 unidentified species were recorded.

Gaur *et al.*, (1998b), have studied the variation in the spore density and percentage of root length of tree species colonized by arbuscular mycorrhizal fungi at a rehabilitated water logged site in Haryana, where water logging was a usual phenomenon that extended for more than nine months annually. Arbuscular mycorrhizal Fungal (AMF) distribution in all the plots represented by various tree species, viz., *Terminalia arjuna, Syzygium cuminii, Populus euphratica* and naturally grown *Typha elephantina* was found to be significantly different. Distribution profile showed the dominance of the two genera, viz., *Glomus* and *Gigaspora*, *Glomus* was dominant in the *T.arjuna* plot while *Gigaspora* was abundant in the *P. euphratica* plot. *Glomus* showed positive correlation to available soil P while *Gigaspora* showed positive correlation to organic matter content. Both the genera also predominated at the naturally grown *T.elephantine* plot.

Mishra *et al.*, (1999), have studied the effect of application of AM fungi in enhancing water stress tolerance capability of pasture legume *Stylosanthes hamata* grown in degraded forest wasteland soil Chandaka, Orissa in presence of Jalashakti (a chemical polymer known for increasing water holding capacity formulated by NCL) under different water regimes (one day and three days gap watering). Application of AM fungi in presence of *Rhizobium* was seen to be more effective in enhancing water stress tolerance capability in *stylosanthes,* indicated through enhancement of plant growth, biomass and nodulation capacity along with rhizopheric microbial activity but to limited extent. Effectiveness of AM fungi inoculation was seen to very with water regime and Jalsakti application. On the contrary, there was indication of retardation of growth and nodulation as well as rhizospheric microbial activity in *Stylosanthes,* with Jalsakti application under adequate watering condition (daily watering) but the effectiveness of AM fungi inoculation stimulating growth and nodulation capacity of *stylosanthes* was seen to enhance significantly under limited water supply condition (3 days gap) in presence of Jalslakti. There was about 66% of water economy along with induction of drought tolerance capacity due to Jalshti application and AM fungi inoculation.

Bharttacharya *et al.*, (1999), have studied the mycorrhizal dependency, phosphorus utilization efficiency and relevance of mycorrhizal for bamboo cultivation in laterite wasteland. Most commercially important bamboo species have a coarse surface spreading root system. Natural root colonization by arbuscular mycorrhiza and inoculum availability in the rhizosphere of most species of bamboo in wasteland soils are reported to be low. *Dendrocalamus strictus*, an important bamboo species in a phosphorus deficient laterite soil showed above 40% Arbuscular mycorrhizal dependency for shoot dry matter production and phosphorus acquisition at 360 days. Phosphorus Utilization Efficiency (PUE) of mycorrhiza inoculated seedling plants was lower than that of uninoculated plants due to a higher net phosphorus acquisition rate than that could be currently by the inoculated plants. Both inoculation efficiency and PUE declined at high phosphorus level but significant response efficiency was available even up to about 25 ppm soils P. Mycorrhiza inoculation at 4 ppm level of soil P showed response equivalence to 130 ppm soil supply. Based on this evidence, the root morphological traits and the reported mycorrhizal relation, bamboo, species, such as *D. strictus* may be considered as suitable for application of AM fungal technology in wasteland plantations.

Bhatia *et al.*, (1999), have assessed the growth of *Prosopis juliflora* and its contribution to soil enrichment, following inoculation with three Arbuscular mycorrhizal isolates, namely *Glomus caledonium, Gigaspora calospora*, and an indigenous strain; and two *Rhizobium* isolates, namely P-5 and Tal-600. The trees were 6-year old, growing on a semi-arid wasteland. Results showed significant improvement in biomass of closely spaced *P. juliflora* inoculated with *G. caledonium* alone. Growth of *P. juliflora* on a relatively nutrient deficient wasteland significantly restored the soil productivity by ameliorating and enriching the soil underneath. A significant reduction in soil reaction (pH) and considerable improvement in soil organic carbon build up and Olsen's-P at both the depths was observed, in all the treatments over zero-time. Moreover, although total nitrogen content increased over control, but the increment was not significant when comparisons were made between respective treatments at zero-time and after six years of growth.

Meshram *et al.*, (1997), have studied the effects of a selected isolate of *Azotobacter chroococcum* (SM) and of the AM fungus *Glomus fasciculatum* on growth and biomass production of *Eucalyptus camadulensis* on waste/barren land at Nagpur, Maharashtra. The field treatment included factorial combinations of biofertilizer (*Azotobacter chroococcum* SM_3 + *Glomus fasciculatum*) with farmyard manure and chemical fertilizer (N as urea and P) and a control treatment. Of the 8 treatment combinations, *A. chroococcum*+mycorrhizas amended with FYM and N + P gave maximum height and biomass production and plant nitrogen content after two years.

Prasad *et al.*, (1997), have studied the influence of AM fungal macro-and micro-nutrients on vegetative propagation of *Dendrocalamus strictus*. Seed of *Dendrocalamus* were collected from the Forest Research Institute, Dehra Dun, and Uttar Pradesh and germinated in mother beds in April-may, 1995. Young seedlings with 2-3 culms pots filled with 8 kg degraded laterite soil, given various treatments, NPK micronutrients (Zn+Mn+B+Cu+Mo), inoculation with AM fungal from a mass culture, inoculation with *Azotobacter* or all possible combinations of these 4 treatments. After 150 days, maximum Culm numbers and height were recorded in the NPK + micronutrients + AM treatment followed by NPK alone, NPK + AM + *Azotobacter* etc.

Bhatia *et al.*, (1998), have assessed the growth of *Prosopis juliflora* and its contribution to soil enrichment following inoculation with three Arbuscular mycorrhizal isolates, *Glomus caledonius*, *Gigaspora calospora*, and an indigenous strain, and two *Rhizobium* isolates, P-5 and Tal-600. The trees were 6 years old and grew on a semi-arid wasteland. There was a significant increase in the biomass of closely spaced *P. juliflora* inoculated with *G. caledonius* alone. *P. juliflora*, growing on a relatively nutrient-deficient wasteland, significantly restored the soil productivity by ameliorating and enriching the soil. A significant reduction in the soil reaction (pH) and a considerable improvement in soil organic carbon build-up and phosphorus, at both depths, were observed in all the treatments by the end of the experiment. Moreover, although the total nitrogen content increased in comparison to controls, this increase was not statistically significant when comparisons were made between respective treatments at the beginning of the experiment and after 6 years growth.

Rao and Tak (2001), have studied the effect of soil inoculation with an arbuscular mycorrhizal (AM) fungus, *Glomus fasciculatum*, isolated from gypsum mine spoil, on dry matter production, nutrient uptake and rhizosphere microbial activity in five tree species (*Acacia ampliceps*, *A. eriopoda*, *Albizia lebbeck*, *Azadirachta indica* and *Calophospermum mopana*) growing in gypsum mine spoil. Dry matter production was significantly enhanced upon inoculation to varying degrees depending on the tree species. The level of root infection increased by 14-36% and spore densities in the rhizosphere increased by 49-81. AM fungal inoculation was resulted in enhanced activities of various soil enzymes (dehydrogenase [oxidoreductase], phosphatases [phosphoric monoester hydrolases] and nitrogenase) in the rhizosphere. Total uptake of many nutrients, but not of potassium and sodium, was significantly enhanced upon AM fungal inoculation. It is concluded that arbuscular mycorrhizal fungi are important symbionts for use in the revegetation of gyspsum mine spoils.

Rao *et al.*, (2002), have investigated arbuscular mycorrhizal dependency of two forest tree species, viz. *Enterolobium saman* [*Samanea saman*] and *Acacia melanoxylon* in eleven coalmine-disturbed soils. The study revealed that the mycorrhizal dependency could be correlated with the degree of disturbance of the soil. Mycorrhizal dependency of *E. saman* in different soils ranged between 23 and 63 percent with minimum dependency in 3 incline soils and maximum in 11A incline soils. Mycorrhizal dependency of *A. melanoxylon* ranged between 30 and 61 with minimum in 3 incline soils and maximum in 7B incline soils. Nodulation of the two tree species seedlings also varied with the type of disturbed soil. The effect of *Glomus fasciculatum* with municipal sewage water on *Bauhinia racemosa* was studied by (Lakshman *et al.*, 2004).

7
Sustainable Management

7.1 Sustainable Management of Natural plant resources for Phytoremediation

Phytoremediation of contaminated sites supports the goal of sustainable development by helping to conserve soil as a resource, bring soil back into beneficial use, preventing the spread of pollution to air and water, and reducing the pressure for development on green or agricultural field sites. Phytoremediation offers the possibility of a cost effective remediation means for a wide range of contaminated sites. It will be most applicable to soil contaminations that not so deep from soil surface, relatively non-leachable, and cover a large area. So far, the processes by which of phytoremediation systems accumulate and degrade contaminants are still poorly understood.

Further, the effects for increasing the scope and efficiency of phytoremediation, and for developing phytoremediation systems for sites contaminated with multi-contaminants are urgently necessary. Although some companies have started their business in phytoremediation, phytoremediation has not been fully commercialized. Further research is still needed, and the priorities on phytoremediation for the future should focus on establishing stable and efficient phytoremediation systems through finding more efficient remediation plants and microbes, monitoring current field trials to obtain thorough understanding, developing microbe-plant combination systems, and using genetic engineering technology.

As phytoremediation need a long period, it has not been fully utilized. Further promotion to the practical application of phytoremediation to removal of contaminated soil needs to establish more effective ways for profitable phytoremediation systems. The use of economic plants such as biofuel crops for utilization and remediation of the contaminated sites would be a reasonable choice, as they can both remediate contaminated soils and produce valuable biomass, which could bring income for the owner of the contaminated site. The impact of AM fungi, pressmud and IAA on Tomato seedlings studied by (Kavatagi and Lakshman, 2011).

Phytoremediation are expected to be used as a vital tool in sustainable management of contaminated soils. Contaminated site managers should consider phytoremediation when evaluating remedial alternatives. The effect of salt and acid stress on wheat seedlings has reported by Bheemareddy and Lakshman, 2011).

7.2 Phytoremediation and Reforestation

Phytoremediation and reforestation may be planned together. If we notice natural forests we find, there are types of plants which dominate depending on soil, rain zone, temperature zone and other geological and geographical conditions. But none of the natural forests are of mono but of mixed type. In a view to achieve noticeable results fast growing plants - *Eucalyptus, Acacia, Bahunia,* etc. are planted as monoculture, the effect of which on the soil and other aspects of environment has never been too encouraging. For future remediation planning, mixing fruit bearing trees of herbal and medicinal repute, bellericas, myrobelum, saponin, ritha, oil seed bearing - mahua, plants of timber value, teak, mehagani, etc. will go a long way. Abundance of fruit tress, indirectly relieve the stress for firewood for home cooking. The use of AM fungal technology for forest nursery seedlings in alkaline soils was discussed by (Lakshman, 2014).

Approach and outlook to bioremediation by different nations are different. With their vast land and other natural resources, USA can afford larger and expensive projects. Aquaculture and phytoremediations are good examples. The Albemarle's Magnolia plant in Arkansas with 54 acres artificial marsh is also a tourist attraction. The Rock reed filter marsh at Dugussa's Theodore, Alaska; Biosphere 2 is other example. Europe is a conglomeration of states very adjacent to each other, highly populated, many are landlocked and natural resources are limited. Japan is having more acute situation than in Europe. Germany and Japan have more holistic approach t this aspect of environmental biotechnology and is likely to come out with some newer pathfinders. European ventures are not very much published and details are not available. A close watch on their ambient air and effluent quality indicate that they are conscious of their responsibilities and do the needful in proper timeframe to check deterioration of the ecosystems at large. The mandatory installation of automobile exhaust converters is a good example. Biotreatment of most gaseous and fluid wastes are major remediation targets of the European communities, is likely to take leadership, A few Japanese achievements may be cited here. Since 1990 Japanese Government's official policy demands that the industries will work but pay attention to the preservation of global environment (not their own alone!) and treat wastes and conserve natural resources. The importance of AM fungal colonization on forest tree species was reported by (Lakshman, 2008; Bheemareddy and Lakshman, 2008).

8

Enzymes used by Plants to Detoxify Organic Compounds

A variety of plant and microbial enzymes are involved in phytoremediation, including nitroreductases, glycosyl and glutathione transferases, oxidases, phosphatases, nitrilases, and dehalogenases. These enzymes are involved in the transformation of toxic xenobiotic compounds such as explosives, pesticides and halogenated organic compounds. Whereas nitroreductases and glycosyltransferases play an important role in the transformation and conjugation of explosives, dehalogenases and many diverse organophosphatases detoxify other contaminants by reducing either halogen groups or organically bound phosphate, respectively. In addition, plants and many microorganisms contain an abundance of oxidases such as laccases and peroxidases. These enzymes are involved in many plant processes including lignification and defense, and may play a role in the metabolism of explosives. Finally, nitrilases catalyze the biodegradation of endogenous plant hormones and commercial herbicides. Knowledge of these metabolic enzymes that catalyze detoxification provides important fundamental insight into metabolic pathways for many organic contaminants and ultimately determines the effectiveness of phytoremediation. Interestingly, the toxicity of dimethoate on primary productivity of Lentic aquatic system was proposed by (Ratageri *et al.*, 2006).

Phytoremediation is a cost effective and nature friendly biotechnology that is powered by microbial enzymes. The research activity in this area would contribute towards developing advanced bioprocess technology to reduce the toxicity of the pollutants and also to obtain novel useful substances. The information on the mechanisms of bioremediation-related enzymes such as oxido-reductases and hydrolases has been extensively studied. A large number of enzymes from bacteria, fungi, and plants have been reported to be involved in the biodegradation of toxic organic pollutants. This topic includes the descriptive information on the enzymes from various microorganisms involved in the biodegradation of wide range of pollutants, applications, and suggestions required to overcome the limitations of their efficient use. The quality of life on the earth is linked inextricably to the overall quality of the environment. Unfortunately the progress in science, technology, and industry a large amount ranging from raw sewage to nuclear waste is let out or dumped into the ecosystem thereby posing a serious problem for survival of mankind itself on earth.

In the past, wastes were traditionally disposed by digging a hole and filling it with waste material. This mode of waste disposal was difficult to sustain owing to lack of new place every time to dump. New technologies for waste disposal that use high-temperature incineration and chemical decomposition (e.g., base-catalyzed dechlorination, UV oxidation) have evolved. Although they can be very effective at reducing wide a range of contaminants but at the same time have several drawbacks. These methods are complex, uneconomical, and lack public acceptance. The associated deficiencies in these methods have focused efforts towards harnessing modern-day bioremediation process as a suitable alternative.

Bioremediation is a microorganism mediated transformation or degradation of contaminants into nonhazardous or less-hazardous substances. The employability of various organisms like bacteria, fungi, algae, and plants for efficient bioremediation of pollutants has been reported (Vidali, 2001; M. Leung, 2004). The involvement of plants in the bioremediation of pollutants is called as phytoremediation. The process of phytoremediation is an emerging green technology that facilitates the removal or degradation of the toxic chemicals in soils, sediments, groundwater, surface water, and air. Genetically, engineered plants are also in use. For instance arsenic is phytoremediated by genetically modified plants such as *Arabidopsis thaliana* which expresses two bacterial genes. One of these genes allows the plant to modify arsenate into arsenite and the second one binds the modified arsenite and stores it in the vacuoles (M. Leung, 2004). The role of microorganisms and their contribution on climate change has been reported by (Channabasava and Lakshman, 2013).

The process of bioremediation mainly depends on microorganisms which enzymatically attack the pollutants and convert them to innocuous products. As bioremediation can be effective only where environmental conditions permit microbial growth and activity, its application often involves the manipulation of environmental parameters to allow microbial growth and degradation to proceed at a faster rate.

The process of bioremediation is a very slow process. Only certain species of bacteria and fungi have proven their ability as potent pollutant degraders. Many strains are known to be effective as bioremediation agents but only under laboratory conditions. The limitation of bacterial growth is under the influence of pH, temperature, oxygen, soil structure, moisture and appropriate level of nutrients, poor bioavailability of contaminants, and presence of other toxic compounds. Although microorganisms can exist in extreme environment, most of them prefer optimal condition a situation that is difficult to achieve outside the laboratory (Vidali, 2001; Bernhard-Reversat and Schwartz, 1997; Dua, Singh, Sethunathan, and Johri, 2002; Dana and Bauder, 2011). Most bioremediation systems operate under aerobic conditions, but anaerobic environments may also permit microbial degradation of recalcitrant molecules. Both bacteria and fungi rely on the participation of different intracellular and extracellular enzymes respectively for the remediation of recalcitrant and lignin and organopollutants (Vidali, 2001; Hammel, 1997).

Enzymes

Enzymes are biological catalysts that facilitate the conversion of substrates into products by providing favorable conditions that lower the activation energy of the reaction. An enzyme may be a protein or a glycoprotein and consists of at least one polypeptide moiety. The regions of the enzyme that are directly involved in the catalytic process are called the active sites. An enzyme may have one or more groups that are essential for catalytic activity associated with the active sites through either covalent or noncovalent bonds; the protein or glycoprotein moiety in such an enzyme is called the apoenzyme, while the nonprotein moiety is called the prosthetic group. The combination of the apoenzyme with the prosthetic group yields the holoenzyme. Enzyme names apply to a single catalytic entity, rather than to a series of individually catalyzed reactions. Names are related to the function of the enzyme, in particular, to the type of reaction catalyzed (Lehninger, Nelson, and Cox, 2004).

Classification of Enzymes

Enzymes are generally grouped into six categories. They are as follows:

1. Oxidoreductases
2. Transferases
3. Hydrolases
4. Lyases
5. Isomerases
6. Ligases (synthetases)

Oxidoreductases catalyze the transfer electrons and protons from a donor to an acceptor. Transferases catalyze the transfer of a functional group from a donor to an acceptor. Hydrolases facilitate the cleavage of C–C, C–O, C–N, and other bonds by water. Lyases catalyze the cleavage of these same bonds by elimination, leaving double bonds (or, in the reverse mode, catalyze the addition of groups across double bonds). Isomerases facilitate geometric or structural rearrangements or isomerizations. Finally, ligases catalyze the joining of two molecules (Lehninger, Nelson, and Cox, 2004).

Phytoremediation Outlook

9.1 Advantages, Disadvantages and Limitations

Phytoremediation has received considerable attention because it offers a cheaper, easier and environmentally sound pollution-remediation option. Seeding a field of plants and harvesting them to extract the pollutant is much cheaper than removing huge amounts of contaminated soil. In cases where it is not economically feasible to recycle the metals, it is still cheaper to dispose of a small amount of contaminated plant mass in a hazardous waste landfill rather than acres of topsoil. Physicochemical technologies for soil remediation render the land useless as they remove all biological activities, including useful microbes, nitrogen fixing bacteria and mycorrhizal-fungi, as well as fauna in the process of decontamination.

Advantages

1. Environment friendly, cost-effective, and aesthetically pleasing.
2. Metals absorbed by the plants may be extracted from harvested plant biomass and then recycled.
3. Phytoremediation can be used to clean up a large variety of contaminants.
4. May reduce the entry of contaminants into the environment by preventing their leakage in to the ground water systems.

Disadvantages

1. Slow process as it relies on natural cycle of plants.
2. Limited by distance, phytoremediation works best when the contamination is within reach of the plant roots, 3-6ft underground for herbaceous plants and 10-15 feet for trees.
3. Some plants absorb a lot of poisonous metals, making them a potential risk to the food chain if animals feed upon them.

Limitations

Phytoremediation could potentially have adverse impacts at the site and surroundings. The potential adverse impacts listed below are not meant to discourage the potential use of phytoremediation, but rather to make the reader aware of potential drawbacks. Possible adverse impacts or disadvantages of phytoremediation include:

1. The introduction and spreading of a potentially undesirable plant (noxious or invasive weeds) that can take over local vegetation must be avoided. Plants should not have an adverse effect on the local ecosystem. The vulnerability of the surrounding area to the selected vegetation and the vegetation's impact must be examined.

2. If plants are used that would contribute an unacceptable amount to local allergen loadings, the plants must be harvested before release of the allergen or treated to decrease the impact. Weed species which release allergens or toxins should not be promoted for Phyto-extraction; hence weeds like *Parthenium histosprous* and *Argemone mexacana* etc. should be avoided in these studies.

3. If soil additives like chelatins and pesticides are used during preparation or maintenance of the system, the impact of these must be carefully monitored, and negative impacts on nearby crop or residential areas should be studied carefully.

4. The animals utilize available food material from plant biomass, thus increasing the chances of introducing the heavy metal/pollutant into the food chain. The plants might attract unwelcome animals that become dependent on the biomass material for food. Hence one of the plant selection criteria should be non-edible nature of the plants.

5. Harvested plant biomass from Phyto-extraction site may be classified as a hazardous waste hence disposal should be proper.

6. Commercial Phyto-extraction has been constrained by the expectation that the site remediation should be achieved in a time comparable to other clean up technologies.

9.2 Plant Disposal Considerations

Due to the growth of vegetation, the mass of plant material will increase with time. Depending on the type of phytoremediation, the biomass that must be removed from the active system will vary. Relatively permanent long-term systems that rely on the establishment of mature vegetation (e.g., poplar trees or grass for rhizodegradation) will not require periodic planned removal of the biomass. In all phytoremediation systems, however, some biomass such as dead or diseased plants, fallen leaves, fallen limbs, or pruned material have to be removed occasionally to maintain good operation of the system. These uncontaminated plant materials will need to be harvested, stored, and disposed of as necessary. It will be important to confirm that the hazardous substances of the plant material do not mix into the ecosystem. If the material could be composted, off-site disposal will be required. When appropriate disposal is an important regulatory concern, the use of lower biomass producing Hyper-accumulator plants the biomass can be used as a source of energy (Moffat, 1995). This will reduce the amount of waste needed to be handled and also open the possibility of biomining reported by Schnoor (1997).

Harvest and Disposal of Plants

When the plants have absorbed and accumulated contaminants, they can be harvested and discarded. If organic chemical contaminants are degraded into molecules like water and carbon dioxide, the plants may not require any special method of disposal. Controlled incineration is the standard method of disposal and then appropriate disposal of the ashes. Recovery of the original metals from the ashes is being investigated.

Sustainable Development and Phytoremediation

Sustainable technologies should be economically viable and environmentally compatible. Phytoremediation can be seen as an effective sustainable technology being more cost effective than traditional remediation methods, less laborious and does not disturb the natural surroundings of the contaminated site. Although this process takes time to gain results, it is one of the best demonstrations of using naturally existing resources to clean up contaminants.

Conclusion

Phytoremediation is a fast developing field, since last 10 years a lot of field applications were initiated all over the world. This sustainable and inexpensive process is fast emerging as a viable alternative to conventional remediation methods and will be most suitable for a developing country like India. To date, very few plants have been described worldwide that fulfills all the phytoremediation criteria. Lot of work is needed and rapidly growing nonaccumultator should be genetically modified so that it achieves some of the properties of hyperaccumulators. Recent progress in determining the molecular basis for metal accumulation and tolerance by hyperaccumulators has been significant, and provides us with a strong basis to outline some strategies for achieving these goals. Meanwhile a lot of work is needed to commercialize this technology in India.

Summary

As stated in Barry Commoner's II law of ecology 'Every thing must go somewhere'. Thus a hazardous pollutant, if not transformed naturally by either chemical or biological reactions, will remain indefinitely in the environment and its total mass will not change with time. In case of organic soil pollution plants can degrade them, but since the heavy metals cannot be degraded, they can be best remediated by a technology that can isolate and concentrate them rather than dilute and disperse them. Conventional remediation techniques utilize substantial amount of fossil or conventional fuel and generate a lot of waste; these wastes act as secondary pollutants and need further disposal. In all, these lead to increase in entropy of the subsystem, and are not taken into account. Both the conservation of mass principle and second law of thermodynamics indicate that most remediation technologies though successfully remediate a specific pollutant, unavoidably cause negative environmental impacts elsewhere. This chapter elaborates the advantages of phytoremediation over other remediation techniques, in view of the power to concentrate or isolate pollutant without excavating or disturbing the site. Phytoremediation seems promising as overall expenses on energy is low and it requires only solar energy to run the process, additionally it provides biomass which can be used as an energy source.

Phytoremediation technology is applicable to a broad range of contaminants, including metals and radionuclides, as well as organic compounds like chlorinated solvents, polycyclic aromatic hydrocarbons, pesticides, explosives and surfactants. The largest barriers to the advancement of phytoremediations, however, may be public opposition to genetic modification in general. Because all natural hyperaccumulator species are small in size, genetic modification can be used to introduce this technology to other species or to increase the biomass of the natural hyperaccumulators in order to create effective phytoremediators. This public opposition was the same fears that surround the issue of genetic modification of crops, and include concerns regarding decreased biodiversity, the entry of potentially harmful genes into products consumed by humans and the slippery slope created by introducing and transferring novel, foreign DNA between non-related species. Nonetheless, the benefits of using phytoremediation to restore balance to a stressed environment seem to far outweigh the costs.

REFERENCES

A Bio-Wise 2003. Contaminated Land Remediation: A Review of Biological Technology, London, and DTI.

Aboulroos, S. A., Helal, M. I. D. and Kamel, M. M. 2006. Remediation of Pb and Cd polluted soils using *in situ* immobilization and phytoextraction techniques. *Soil Sediment Contam.* 15: 199-215.

Ahalya, N.; Ramachandra, T.V.; Kanamadi, R.D. (December 2003). "Biosorption of Heavy Metals". *Research Journal of Chemistry and Environment* 7 (4).

Alder, T. 1996. Botanical cleanup crews. *Sci. News.* 150: pp. 42-43.

Alleman, James E. and Prakasam, T.B.S. 1983. "Reflections on Seven Decades of Activated Sludge History". *Water Pollution Control Federation.* 55(5): pp. 436–443.

Alloway, B. J. (1990). "Cadmium, Heavy Metal in Soil. London," Blackje and Son, London.

Amir, H. and Pineau, R. 2003. Release of Ni and Co by microbial activity in New Caledonian ultramafic soils. *Can. J. Microbiol.* 49: pp- 288-293.

Andrade, J. C. M. and Mahler, C. F. 2002. Soil Phytoremediation. In 4th International Conference on Engineering Geotechnology, Riode Janeioro, Brazil.

Antonovies, J., Bradhave, A. D. and Turner, R. G. 1971. Heavy Metal tolerance in plants. *Adv. Ecol. Res.,* 75: 1-85.

Arundati Pal and Paul, A.K. 2007. Rhizosphere of Nickel hyper accumulating plants: a niche for Nickel – Resistant Bacteria. *Rhizosphere Biotechnology: plant growth retrospect and prospect.* Published by A.K. Roy. pp. 121-134.

Baker, A. J. M. and Brooks, R. R. 1989. Terrestrial higher plants which hyper-accumulate metallic elements. A review of their distribution, ecology and phytochemistry. *Biorecovery.* 1: pp. 81-126.

Baker, A. J. M. and Walker, P. L. 1990. Ecophysiology of metal uptake by tolerant plants. In: A. J. Shaw (ed.). *Heavy metal tolerance in plants: Evolutionary aspects* CRC Press Inc., Boca Raton. FL.

Barea, J. M., Azcon Anguilar, C. 1982. Production of plant growth regulating substances by VAM Fungus, *Glomus mossae*. *Appl. Environ. Microbiol.* 43: 810-813.

Basu, M., Bhattacharya, S. and Paul, A.K. 1997. Isolation and characterization of chromium-resistant bacteria from tannery effluents. *Bull. Environ. Contom. Taxicol.* 58: pp. 535-542.

Bernhard-Reversat, F and Schwartz, D. 1997. "Change in lignin content during litter decomposition in tropical forest soils (Congo): comparison of exotic plantations and native stands," Comptes Rendus de l'Academie de Sciences—Serie IIa, vol. 325, no. 6, pp. 427–432.

Berti, W. R. and Cunningham, S. D. 2000. Phytostabilization of Metals. In: Ruskin I & Ensley BD (eds), *Phytoremediation of toxic metals; Using plants to clean up the environment*. Wiley, New York. pp. 71-88.

Bhatia, N.P., Adholeya, A. and Sharma, A. 1999. Inoculation effects of AM fungi and *Rhizobium* spp. On growth performance of *Prosopis juliflora* and consequent improvement in productivity of semi-arid soil. Abstracts: *National Conference of Mycorrhiza*, Barkatulla, University, Bhopal: pp. 63.

Bhatia, N. P., Adholeya, A. and Sharma, A. 1998. Biomass production and changes in soil productivity during long- term cultivation of *Prosopis juliflora* (Swartz) DC. inoculated with VA mycorrhiza and *Rhizobium* spp. in a semi-arid wasteland Biological Fertility of Soils. 26: pp. 208-214.

Bhattacharya, P.M., Misra, D., Saha, F. and chaudhuri, S. 2000. Arbuscularmycorrhizal dependency of *Eucalyptus tereticornis* Sm: how real is it? *Mycological News.* 12(3): pp. 11-15.

Bhattacharya, P.M., Misra, D., Saha, J. and Chaudhuri, S. 1999. Mycorrhiza dependency, phosporus utilization efficiency and relevance of mycorrhiza for bamboo cultivation in laterite wasteland. Abstracts: *National Conference on Mycorrhiza*, Barkatulla University, Bhopal: p. 47.

Bhatti, Haq N., Nasir, Abdul W., Hanif, Muhammad. A, 2010. Efficacy of *Daucus carota* L. waste biomass for the removal of chromium from aqueous solutions. *Desalination* Vol. 253 (1–3): pp. 78–87

Bheemareddy, V. S. and Lakshman, H. C. 2008. Survey of VAM Fungi in some Rare, Endangered and Threatened Tree Species. Forest Biodiversity, Eds. K. Muthuchelian, S. Kannaiyan & A. Gopalam. Associated Publishing Co., New Delhi. 2: pp. 231-239.

Bheemareddy, V.S. and Lakshman, H.C. 2011. Effect of salt and acid stress on *Triticum aestivum* L. inoculated with *Glomus fasiculatum*. An international *Journal of Agricultural Technology.* 7(4): pp. 945-956.

Bisen, P.S., Gour, R.K. Jain, R.K., Dev, A. and Sengupta, L.K. (1996). VAM colonization in tree species planted in Cu, Al and coal mines of Madhya Pradesh with special reference to *Glomus mosseae*. *Mycorrhiza News*. 8(1):9-11.

Blaylock, M. J. and Huang, J. W. 2000. Phytoextraction of Heavy Metals. In: Raskin, I. and Ensley, B. D. (eds), Phytoremediation toxic metals: Using Plants to Clean Up the Environment, New York, John Wiley and Sons. pp. 53-69.

Bowen, G. C. and Rovira A. D. 1991. The Rhizosphere – the hidden half of the hidden half. In the Roots- the hidden half, eds. Y Waisel, A Eshel, U Kaffkafi, NY Marcel Dekker. pp.641-649.

Bradley, R., Burt, A. J. and Read, D. J. 1981. Mycorrhizal infection and resistance to heavy metal toxicity in *Calluna vulgaris*. *Nature*, 292: pp. 235-237.

Brooks, R.R. 1987. Serpentine and its vegetation, amulti-disciplinary approach. Chapter 3, Croom Helm, London, pp. 18-31.

Brown, S. L., Chaney, R. L., Angle, J. S. and Baker, A. J. M. 1994. Phytoremediation potential of *Thlaspi caerulescens* and bladder campion for zinc and cadmium contaminated soil. *J. Environ. Qual.* 23: pp. 1151-1157.

Brown, M. T. and Wilkins, D. A. 1985a. Zinc tolerance of *Amanita* and *Paxillus*. *Trans. Brit. Mycol. Soc.*, 84: pp. 367-369.

Brown, M. T. and Wilkins, D. A. 1985a. Zinc tolerance of *Betula*. *New Phytol.*, 99: pp. 101-106.

Burrows, W. D. 1982. Tertiary treatment of effluent from holston army ammunition plant. US army armament R&D Command, dover, NJ, report no. 8207.

Cataldo, D. A. 1990. An evaluation of the environmental fate and behavior of mountains materials (TND. RDX) in soil and plant system: RDX, U. S. army biomedical research and development laboratory.

Chandra, K. K., Chaturvedi, P. and Jamaluddin. 1997. Comparative study of VA mycorrhizal root infection and spore density in *Grevilia pteridifolia* in undisturbed and disturbed area.

Chandra, S. Dubey, R. and Kehri, H.K. 2000b. Wasteland management viapetroleum plantation and VAM technology. In: *Innovative Approaches inMicrobiology*. (Eds. D.K. Maheshwari and R.C Dubey), Bishen Singh, Mahendra Pal Singh, Dehra Dun: pp. 75-97.

Chen, D. 1993. Plant uptake and absorption of RDX agronomy, urbana- champaign, University of illiois. pp. 103.

Chaney, R. L., Li, Y. M., Angle, J. S., Baker, A. J. M., Reeves, R. D., Brown, S. L., Homer, F. A., Malik, M. and Chin, M. 1999. Improving metal hyperaccumulators wild plants to develop commercial Phyto-extraction systems: approaches and progress. In: (ed. N Terry. GS Banuelos), *Phytoremediation of contaminated soil and water*. CRC press.

Channabasava, A and Lakshman, H. C. 2012. AM fungi and mine spoil consortium: a microbial approach for enhancing Proso millet biomass and yield. *Inter. J. of Pharma and Biosci*. 3(4): pp. 676-684.

Channabasava, A., Lakshman, H. C. and Kavatagi, P. K. 2013. Biogeography of microorganisms and their contribution to climate change. *Research Arena*. Vol. 1(9): pp. 81-84.

Cheng, S. 2012. "Heavy Metals in Plants and Phytoremediation. "http://ir.ihb. ac.cn/bitstream/152342/9658/1/Heavy%20metals%20in%%20journals.pdf.

Cheraghi, M., Lorestani, B. Khorasani, N., Yousefi, N. and Karami, M. 2011. Findings on the phytoextraction and phytostabilization of soils contaminated with Heavy Metals. *Boil. Trace. Elem. Res*. 144: pp. 1133-1141.

Chiba, S. and Takahashi, K. 1977. Studies on Heavy Metal pollution in agricultural land (2). Absorption of Cadmium & growth retardation in forage crops. *Bull. Shikoku Agri. Exp. Station.*, 30: pp. 49-73.

Chigho, F. E., Smith, R. W. and Shore, F. L. 1982. Uptake of Arsenic, Cadmium, Lead and Mercury from polluted waters by the water hyacinth (*Eichhornia crassipes*). *Environ. Pollut.* (Ser.B), 27: pp.31-36.

Cholpecka, A., Bacon, J.R., Wilson and M. J. and Kay, J. 1996. "Heavy Metals in the Environment. Forms of Cadmium, Lead and Zinc in Contaminated Soils from Southwest Poland," *Journal of Environmental Quality*, 25 (1) pp.69-79.

Clemens, S., Palmgren and Kramer, U. 2002. "A Long Way Ahead: Understanding and Engineering Plant Metal Accumulation," *Trends in Plant Science*. 7(7): pp. 309-314.

Colpaert, J. V. and Van Assche, J. A. 1992a. Zinc toxicity in ectomycorrhizal *Pinus sylvestris*. *Plant Soil*. 143: pp. 201-211.

Colpaert, J. V. and Van Assche, J. A. 1992b. The effects of cadmium and the cadmium-zinc interaction on the axenic growth of ectomycorrhizal fungi. *Plant Soil*. 145: pp. 237-243.

Conesa, H. M., Evangelou, M. W. H., Robinson, B. H. and Schulin, R. 2012. A critical view of Current state of phytotechnologies to remediate soils: still a promising tool?

Cortez, P. C. C. 2005. "Assessment and Phytoremediation of Heavy Metals in the Panlasian Creek," Science City of Munoz, Nueva Eeija, Philippines.

Cumming, J. R. and Weinstein, L. H. 1990. Aluminium mycorrhizal interactions in the physiology of pitch pine seedlings. *Plant Soil*. 125: pp. 7-18.

Cunningham, S.D., Berti, W. R. and Huang, J. W. 1995. Phytomediation of contaminated soils. *TIBTECH*, 13: 393-397.

Cunningham, S.D., Huang, J. W., Chen, J. and Berti, W. R. 1996. *Abstracts of papers of the American Chemical Society*. 212: .pp. 87.

Dana, L. D. and Bauder, J. W. 2011. A General Essay on Bioremediation of Contaminated Soil, Montana State University, Bozeman, Mont, USA.

D. K. Purohit, S. Ram Reddy, M.A. Singara Charya, S. Girisham.): 289-293. Sengupta, A. and Chaudhuri, S. 1995. Effect of dual inoculation of Rhizobium and mycorrhiza on growth response of *Sesbania grandiflora*. L. in coastal saline And sand dune soil. *Indian Journal of Forestry*. 18(1): pp. 35-37.

De, K. B. 2008. Bioremediation-Phytoremediation and Reforestation. Textbook of Agricultural Biotechnology. Edited by Ahindra Nag. Published by Asoke k. Ghosh. PHI Learning Private Limited. New Delhi. pp. 151.

Dieberg, F. E., Debus, T. A. and Goulet, N. A. 1987. Removal of copper and lead using a thin film technique. In Reddy, K. R. and Smith, W. H. (eds) Aquatic plants for water treatment and resource recovery. Magnolia Publishing. New York. pp. 497-504.

Dietz, A. C. and Schnoor, J. L. 2001. Advances in phytoremediation. *Enviro health Perspective*. 109: 63-168.

Dixon, R. K. and Buschena, C. A. 1988. Response of ectomycorrhizal *Pinus banksiana* and *Picea glauca* to heavy metal in soil. *Plant Soil*. 105:265-271.

Dixon, R. K., Mukerji, K. G., Chamola, B. P. and Kaushik, A. 1997. Vesicular Arbuscular mycorrhizal symbiosis in relation to forestation in arid land. *Annals of Forestry*. 5 (1): pp. 1-9.

Dobson, M. C. and Moffat, A. J. 1995. A re-evaluation of objection to tree planting on containment landfills. *Waste Management and Research*. 13(6): p. 579.

Donald, J. and Freeman, P. E. 1991. Applicaton of the powdered activated carbon / activated carbon / activated sludge *PACT* and Wet air oxidation process to the treatment of explosives contaminated waste water, *proceedings of the ninth international symposium on compatibility of plastics and other materials with explosives, propellants*, San Diego.

Doss, D. D., Bagyaraj, D. J. and Syamsunder, J. 1988. Anatomical and histochemical changes in the roots and leaves of finger millet colonized by VA mycorrhiza. *Indian Journal of Microbiology*. 28(3-4): pp. 276-280.

Dua, M., Singh, A., Sethunathan, N and Johri, A. 2002. "Biotechnology and bioremediation: successes and limitations," *Applied Microbiology and Biotechnology*, vol. 59, no. 2-3, pp. 143–152,

Dushenkov, S., Vasudev, D., Kapulnic, Y., Gleba, D., Fleisher, D., Ting, K. C. and Ensley, B. 1997. Removal of Uranium from water using terrestrial plants. *Environ. Sci. Technol.* 31: pp.3468-3474.

Dushenkov, V., Nanda Kumar, P. B. A., Motto, H. and Ruskin, I. 1995. Rhizofiltration-the use of plants to remove Heavy Metals from aqueous streams. *Environ. Sci. Technol.* 29: pp. 1239-1245.

Eapen, S., Singh, S., Thorat, V., Kaushik, C. P., Kanwar, R. and D' Souza, S. F. 2007. Phytoremediation of radiostrontium (^{90}Sr) and radiocesium (^{137}Cs) using giant milky weed (*Calotropis gigantea* R.Br.) plants. *Chemosphere*, 65(11): pp. 2071-3.

Ebbs, S. D. and Kochian, L. V. 1997. Toxicity of zinc and copper to *Brassica* species: Implications for phytoremediation. *J. Environ. Qual.* 26: pp. 776-781.

El-Saidi, M. T., Hegazy, W. A., Kortam, M. and El-Zeiny, H. 1983. The use of saline water for propagation and its effects on growth and yield of cotton plants. In: *Proc.1st conf Agric. Egyptian Soc. Crop. Sci.* pp. 739-753.

Entry, J. A., Cromack, K. Jr., Stafford, S. G. and Castellano, M. A. 1987. The effect of pH and aluminium concentration on ectomycorrhizal formation in *Abies balsamea. Can. J. For. Res.* 17: pp. 865-871.

Fletcher, J. S. and Hegde, R. S. 1995. Release of phenols by perrennial plant roots and their potential importance in bioremediation. *Chemosphere.* 31: pp. 3009-3016.

Fulekar, M. H., Singh, A., Thorat, V. Kaushik, C. P. and Eapen, S. 2010. Phytoremediation of ^{137}Cs from low level nuclear waste using *Catharanthus roseus. Indian J. Pure. Appl. Phys.* 48: pp. 516-519.

Gadd, G. M. 2000. Bioremedial potential of microbial mechanisms of metal mobilization and immobilization. *Curr. Opi. Biotech.* 11: pp. 271-279.

Gadd, G. M. and White, C. 1993. Microbial treatment of metal Pollution – a working biotechnology? *Tibtech.* 11: pp. 353.

Galli, U., Schuepp, H. and Brunold, C. 1994. Heavy metal binding by mycorrhizal fungi. *Physiol. Planta.* 92: pp. 364-368.

Garbasu, C. 2002. Phytoremediation: A Technology Using Green Plants to Remove Contaminants from Polluted Areas," *Reviews on Environmental Health*, Vol. 17, No.3, pp.173-188.

Gatliff. E.G. 1994. Vegetative remediation process offers advantages over traditional pump-and-treat technologies. *Remed. Summer.* 4(3): 343-352.

Gaur, A., Sharma, M.P. and Adholeya. A 1998b. Variation in the spore density and percentage of root length of tree species colonized by arbuscular mycorrhizal fungi at a rehabilitated waterlogged site. *Journal of Tropical Forest Science.* 10(4): pp. 542-551.

Ghosh, M. and Singh, S. P. 2005. A review on Phytoremediation of Heavy Metals and Utilization of its by products. *Appl. Ecol. Environ. Res.* 3: pp. 1-18.

Gildon, A. and Tinker, P. B. 1981. A Heavy metal tolerant strain of mycorrhizal fungus. *Trans. Brit. Myco. Soc.* 77: pp. 648-649.

Gordon, M. P. 1997. Phytoremediation of chlorinated solvents – poplars remove chlorinated solvents from soil in field trials. *IBS's second annual conference on phytoremediation.* Seattle. WA. International business communications, Southborough. MA.

Gupta, V., Sanjeevkumar and Satyanarayana. 2002. Heavy Metal tolerance of Ectomycorrhzal Fungi and utility of metal tolerant EM Fungal strains in reclaiming waste lands created by Mining operations. *Bioinoculants for Sustainable Agriculture and Forestry.* Chapter-4. pp.27-29.

Hall, I. R. 1977. Species and Mycorrhizal infections of News Zealand Endogonaceae. Trans Br. Mycol. Soc. 68: pp. 341-356.

K. E. Hammel, 1977. " Fungal degradation of lignin," in Driven by Nature: Plant Litter Quality and Decomposition, G. Cadisch and K. E. Giller, Eds., pp. 33–45, CAB International, Wallingford, UK.

Hannink, N., Roser, S. J., French, C. E., Basran, A., Murray, J. S. H., Nicklin, S. and Bruce, N. C. 2001. Phytoremediation of TNT by transgenic plants expressing a bacterial nitroreductase. *Nat. Biotech.* 19: pp. 1108-1172.

Harley, J. L. 1978. Ectomycorrhzas as nutrient absorbing organs. *Proc. Royl. Soc. Lond.* B 203: pp.1-21.

Harley, J. L. and Smith, 1983. Mycorrhizal Symbiosis, Academic Press, New York.

Hartely, J., Cairney, J. W. G. and Meharg, A. A. 1997. Do Ectomycorrhzal fungi exhibit adaptive tolerance to potentially toxic metals in the environment? *Plant Soil.* 189: 303-319.

Hartely, J., Cairney, J. W. G. and Meharg, A. A. 1998. Differential sensitivity of Ectomycorrhzae and their hosts to toxic concentrations of metals. *Proceedings of the Eighth International Symposium on Microbial Ecology.* pp.171. Halifax, Canada.

Harms, H., Bokern, M. Kolb, M. and Bock, C. 2003. Transformation of organic Contaminant by different plant systems. In phytoremediation: *transformation and Control of contaminants.* Ed. McCutcheon SC, Schnoor JL, NY Wiley, pp. 285-316.

Hashem, A. R. 1995. The role of myccorrhizal infection in the resistance of *Vaccinium macrocarpon* to manganese. *Mycorrhiza,* 5: pp. 289-891.

Heale, E. L. and Ormrod, D. P. 1983. Effect of nickel and copper on seed germination, growth and development of seedlings of *Acer pinnata, Betula papyrifera, Picea abies and Pinus banksiena.* Reclamation and Revegetation Research. Vol. 2: pp. 41-54.

Henry, J. R. 2000. In: An overview of the phytoremediation of Lead and Mercury. NNEMS report. Washington, D.C. pp. 3-9.

Hinesly, T. D., Alexander, D. E., Ziegler, E. L. and Barrett, G. L.1978. Zinc and Cd accumulation by corn inbreds grown on sludge amended soil. *Agron. J.* 70: pp. 425-428.

Hirrel, M. C. and Gerdemann, J. W. 1980. Improved growth of onion and bell pepper in saline soils by two vesicular mycorrhizal fungi. *Soil Sci. Amer. J.* 44: pp. 654-655.

Hooker, R. P. and Shoemakerm, C. A. 1985. Aluminium mobilization in an acidic heat water stream; temporal variation and mineral dissolution disequilibria. *Science.* 229: pp. 463-465.

Huang, J. W. and Cunningham, S. D. 1996. "Lead Phytoextraction: Species Variation in Lead Uptake and Translocation," *The News Physiologists,* 134 (1) pp. 302-303.

Huang, J. W., Cunningham, S. D., Chen, J. and Berti, W. R. (1997). "Phytoremediation of Lead-Contaminated Soils: Role of synthetic Chelates in Lead Phytoextraction," *Environmental Science and Technology,* 31(3): pp. 800-805.

Hughes, M. N. and Poole, R. K. 1989. Metals and microorganisms. Chapman and Hall, New York, pp.1-411.

Huhle, B., Heilmeier, H., and Merkel, B. 2008. Potential of *Brassica juncea* and *Helianthus annuus* in Phytoremediation for Uranium, In: *Uranium Min. Hydrogeol.* pp. 307-318.

Hutchinson ,S. L., Schwab, A. P. and Banks, M. K. 2003. Biodegradation of petroleum hydrocarbon in the rhizosphere. In: *Phytoremediation: transformation and control of contaminants*, ed. McCutcheon SC, Schnoor JL, NY Wiley. pp.355-386.

Huttermann, A. 1985. The effects of acid deposition on physiology of the forest ecosystem. *Experientia*, 41: pp. 585-590.

Idris, R., Trifonova, R., Puschenrieter, M., Wenzel, W.W. and Sessitch, A. 2004. Bacterial communities associated with flowering plants of the Ni hyperaccumulator *Thlaspi goesingense. Appl. Environ. Microbiol.* 70: pp. 2667-2677.

Jadia, C. D., Fulekar, M. H. 2008. Phytotoxicity and remediation of Heavy Metals by Fibrous root grass (Sorbhum). *J. Appl. Biosci.* (10): pp. 491-499.

James, B. R. 2001. Remediation-by-reduction strategies for Chromate contaminated soils. *Environ. Geochem. Health.* 23: pp. 175-189.

Johnson, E. R. R. L. and Shilling, D. G. 2002. "Cogon Grass" Plant Conservation.

Kirchner, A. 2001. Mine-Land Restoration: Phytoremediation of Heavy-Metal Contaminated Sites – A Critical View," International Ecological Engineering Society. http://www.iees.ch/EcoEng011/ EcoEng011_R2.html.

Karkhanis, M., Jadia, C. D. and Fulekar, M. H. 2005. Rhizofiltration of metals from coal ash leachate. *Asian. J. Water Environ. Pollut.* 3: pp. 91-94.

Khan, A. G. 1974. The occurrence of mycorrhizas in halophytes, hydrophytes, xerophytes and *Endogone* spores in saline soils, *J. Gen. Microbiol* 81: pp. 7-14.

Kamaruzzaman, B. Y., Ong, K. C. A., Jalal, S., Shabbudin and Mohd Nor, O. 2011. "Accumulation of Lead and Copper in *Rhizophora apiculata* from Setiu Mangrove Forest, Terrengganu, Malaysia."

www.jeb.co.in/journal_issues/200909_sep09_supp/paper_08.pdf

Karenlamp, S., Schat, H., Vangronsveld, J., Verkleij, J. A. C., Van Der Lelie, D., Mergeay, M. and Tervahauta, A. I. 2000. Genetic Engineering in the improvement of plants for Phytoremediation of Metal polluted Soil. *Environ. Pollut.* 107: pp. 225-231.

Kasuya, M. C. M., Muchovej. R. M. C. and Muchovej, J. J. 1990. Influence of aluminium on in vitro formation on *Pinus caribaea* mycorrhizae. *Plant Soil.* 124: pp. 43-77.

Kathiresan, K. and Bingham, B. L. 2011. "Biology of Mangroves and Mangrove Ecosystems."http://faculty.wwu.edu/bingham/ mangroves.pdf

Killham, K. and Firestone, M. K. 1983. Vesicular arbuscular mycorrhizal mediation of grass response to acid and heavy metal deposition. *Plant Soil.* 72: pp. 39-48.

Kothari, I.L., Udhyaya J. and Patel, C.R. 1997. Mycorrhizal biodiversity of petroeffluent-irrigated fields. *Journal of Mycology and Plant Pathology.* 27(3): pp. 275-278.

Kiran.P.Kolkar and Lakshman, H.C. 2010. Use of AMF Inoculation Under different concentration of domestic sewage water to improve growth of *Impatieus balsaminia* L. *International Journal of Agricultural Science.* 6(1): pp. 262-264.

Kramer, U. 2005. Phytoremediation: novel approaches to cleaning up polluted soils. *Curr. Open. Biotechnol.* 16: pp. 133-141.

Krishna H. Waddar and Lakshman, H. C. 2010. Effect of AM Fungi on the Forest tree seedlings of *Tamarindus indica* L. and *Azadirachta indica* Juss. For Integrated Nursery Stock. *Inter. J. of Plant Protection.* 3(2): pp. 248-252.

Kruckeberg, A.R. 1992. Plant life of western North American ultramafics. In *The ecology of areas with serpentinized rocks. A world view.* (Roberts, B.A. and Proctor, J. eds.) kluwer Academic Publishers, Netherlands, pp. 31-73.

Kumar, N. J. I., Soni, H., Kumar, R. N. and Bhatt, I. 2008. Macrophytes in Phytoremediation of Heavy Metal contaminated water and sediments in Pariyez Community Reserve, Gujarat, India. *Turk. J. Fish. Aquat. Sci.* 8: pp. 193- 200.

Lasat, M. M., Ebbs, S. D. and Kochian, L. V. 1998. Phytoextraction of a radiocaesium-contaminated soil: Evaluation of Caesium-137 bioaccumulation in shoots of three plant species. *J.Environ. Qual.* 27: pp. 165-169.

Lasat, M. M. 2000. "Phytoextraction of Metals from Contaminated Soil: A Review of Plant/Soil/Metal Interaction and Assessment of Pertinant Agronomic Issues," Journal of Hazardous Substance Research, Vol. 2, No pp. 1-25.

Lakshman, H. C., Mulla, F. I., Inchal, R. F. and Somvasalu, Y. 2001. Prevalence of arbuscular mycorrhizal colonization in some disputed plants. *Mycorrhiza News.* 13(3): pp. 16-22.

Lakshman, H. C. 2006. Importance of biofertilizers for young seedlings. *Vignana.* 2: pp. 183-186.

Lakshman, H. C. 2008. VAM Fungal Diversity of Forest Tree species in Dry Deciduous Forests of Dharwad. Forest Biodiversity, Eds. K. Muthuchelian, S. Kannaiyan & A. Gopalam. Associated Publishing Co., New Delhi. 2: pp. 223-230.

Lakshman, H. C. 1994. Occurrence of VA-Mycorrhiza in some tropical hydrophytic and xerophytic plants of Dharwad District Karnatak. *J. of Nat. Con.* 6(1): pp. 29-35.

Lakshman, H. C. 1997. Occurrence of VA-Mycorrhiza in some tropical herbs/ shrubs in stock piled mined soils and possible way to revegetation by introducing VAM plants. S.D.M.C.E.T. Dharwad. Spl. Pub. *Environ. Pollution.* 1: pp. 213-221.

Lakshman, H. C. 1998. Interactions between VAM and other beneficial microorganisms and their effect on growth of *Pterocarpus marsupium* Roxb. *Bios.* 23(1 and 2): 49- 56.

Lakshman, H. C. 1999. Aggregation and sand-dune soil by arbuscular mycorrhizal fungi its use in revegetation practices. *Indian J. Environ. Ecoplanning.* 2(3): pp. 247-252.

Lakshman, H. C. 1999. VA-mycorrhizal survey of plant species colonizing beach ponds Karwar in South India. *Asian J. Microbiology, Biotecnology and Environmetal Sciences.* 1(1&2): pp.15-21.

Lakshman, H. C. 2000. Occurrence and tolerance of VAM plants growing on polluted soils with sewage and industrial effluents. *J. of Nat. Con.* 12 (1): pp. 9-18.

Lakshman, H. C. and Inchal, R. F. 2000. Amendment of sterilized mined soils and forest soil on *Cassia* species colonized with VA-mycorrhiza. *J. Nat. Con.* 12(2): pp. 227-236.

Lakshman, H. C. and Hosamani, P. A. 2003. Effect of AM Fungus (*Glomus mosseae*) and municipal sewage water on growth of *Clitoria ternata* (Butterfly pea). *Ecol. Env. and Conservation.* 3(2): pp. 355-358.

Lakshman, H. C. Jayashankar, M. and Kiran P.Kolkar. 2004. Effect of *Glomus fasciculatum* (AMF) and municipal sewage water on growth of *Bauhinia racemosa* Lam. *Indian J. Environ. and Ecoplan* . 8(2): pp. 309-313.

Lakshman, H. C. and Channabasava A. 2013. Mycorrhizoremediation of mine spoil by using foxtail millet inoculated with *Rhizophagus fasciculatus*: an *ex-situ* waste management. *International journal of current science.* 8: pp. 85-92.

Lakshman, H.C. 2014. Arbuscular mycorrhiza fungal association in some trees growing in alkaline soils of Gadag district-Karnataka. *Research Scholar.* 4(1): pp. 16-19.

Lambert, D. H., Baker, D. E. and Cole, H. J. 1979. The role of mycorrhiza in the interactions of phosphorus with zinc, copper and other elements. *Soil S. Amer. J.* 43: 976-980.

Leblova, S., Mucha, A. and Spihauzlova, E. 1986. Compartmentation of Cadmium and Zinc in seedlings of maize (*Zea mays* L.) and induction of metallothion. *Biol. Czechos.,* 4: pp. 77-88.

Lehninger, A. L., Nelson, D. L. and Cox, M. M. 2004. Lehninger's Principles of Biochemistry, W.H. Freeman, New York, NY, USA, 4th edition

Leigh, M. B., Fletcher, J. S., Fu, X. and Schmitz, F. J. 2002. Root turnover: an important source of microbial substances in rhizosphere remediation of recalcitrant contaminants. *Enviro Sci. Techn.* 36: pp. 579-1583.

Lesmana Sisca O., Febriana, Novie., Soetaredjo, Felycia E., Sunarso, Jaka., Ismadji, Suryadi. 2009. "Milestones in the Development of the Activated Sludge Process". *Water Pollution Control Federation.* 37(2): pp. 151–162.

Lesmana, Sisca O., Febriana, Novie., Soetaredjo, Felycia E., Sunarso, Jaka., Ismadji, Suryadi. 2009. "Studies on potential applications of biomass for the separation of heavy metals from water and wastewater". *Biochemical Engineering Journal..* 44(1): pp. 19–41.

Leung, M. 2004. "Bioremediation: techniques for cleaning up a mess," *Journal of Biotechnology,* 2: pp. 18–22.

Li, X. L., Marschner, H. and George, E. 1991. Acquisition of phosphorus and copper by VA-mycorrhizal hyphae and root-to-shoot transport in white clover. *Plant Soil*. 136: pp. 49-57.

Li, Y. M., Chaney, R. Brewer, E., Roseberg, R. Angle, S. J. 2003. Development of a technology for commercial phytoextraction of nickel: economic and technical Considerations. *Plant soil* 249: pp. 107-115.

Lo Buglio, K. F. and Wilcox, H. E. 1988. Growth and survival of ectomycorrhizal and ectendomycorrhizal seedlings of *Pinus resinoea* on iron tailings. *Can. J. Bot.* 66: pp. 55-60.

Macek, T. 2004. Phytoremediation: Biological Cleaning of a Polluted Environment," *Reviews on Environmental Health*. 19(1): pp. 63-82.

Macfarlane, G. R. and Burchett, M. D. 2011. "Toxicity, Growth and Accumulation Relationship of Copper, Lead and Zinc in the Grey Mangrove *Avicennia marina* (Forsk.) Vierh."http:www.mendeley.com/research/toxicity-growth-and-accumalationrelationships-of-copper-lead-and-zinc-in-the-grey-mangrove-avicennia-marina- forsk-vierh/

Ma, L. Q., Komar, K. M. and Tu, C. 2001. A fern that accumulates arsenic. *Nature*. pp. 409:579.

Mandal, B. S. and Kaushik. J. C 1995. Interaction between VA mycorrhizal fungi and *Rhizobium* and their effect on the growth parameters of *Acacia nilotica* (L).Willd ex. Del.*Haryana Agricultural University Journal of Research*. 25(3): pp. 107-111.

Mandal, B.S., Kaushik, J.C. and Singh, R.R. (1995). Relative efficacy of single and multiple inocula of some VAM fungi for *Acacia nilotica* (L). willd ex Del. *Crop Research (Hisar)*. 9(3): pp. 437- 440.

Mannetje,L.2004."Papsalum Conjugatum," http://www.fao.org/ag/AGP/AGP/doc/GBASE/DATA/PF000492.HTM

Mangkoedihardjo, S. and Surahmaida, 2008. Jatropha curcas. L. for Phytoremediation of Lead and Cadmium Polluted Soil. *World Appl. Sc.*: J. 4: pp. 519-522.

Marschner, P., Godbold, D. L. and Jentschke, G. 1996. Dynamics of lead accumulation in mycorrhizal and non mycorrhizal Norway spruce (*Picea abies* (L.) Karst). *Plant Soil*. 178: pp. 239-245.

Mathur, N., Singh, J., Bohra, S., Bohra, A., Mehboob, Vyas, M. and Vyas, A. 2010. Phytoremediation Potential of Some Multipurpose Tree Species of Indian Thar Desert in Oil Contaminated Soil. *Adv. Environ. Biol.* 4(2): pp. 131-137.

Means, J. L. and Hinchee, R. E. 1994. Emerging technology for bioremediation of metals. Lewis publishers, Boca Raton, FL.

Mehrotra, V. S. 1996. Use of revegetated coal mine spoil as source of arbuscular mycorrhizal inoculum for nursery inoculation. *Current Science* 70(1): pp. 73-77.

Mekonnen, A. 2000. "Handbook on Vetiver Grass Technology- From Propagation to Utilization," GTZ IFSP S/GONDER, Ethiopia.

Meshram, S. U., Peshwe, S.A. Joshi, S.N. and Dongre, A.B. 1997. Response of (to) Biofertilizer on (of) biomass production of *Eucalyptus camaldulensis*. *Annals of Forestry*. 5(1): pp. 43-49.

Mishra, A.K. 1994. Strategies for reclamation of mine waste areas through the help of selected microbes: problems and prospects. In: *Crop Epidemics, Microbes and Ecosystem Conservation*. (Eds. G.P Agarwal and S.K. Hasija), Narendra Publishing House, Delhi: pp. 177-184.

Mishra, A.K. Thatio, H.W and Millik, D. 1999. Induction of drought toterance in *Stylosanthes hamata* (L). Taub. With AM fungi and Jalaskti growth in forest degraded wastes of Chandaka in Orissa. Abstracts: *National Conference on Mycorrhiza*, Barkatulla Universtiy Bhopal: pp. 11.

Mishra, O.R., U.S., Sharma, R.A. and Rajput, A.M. 1995b. Response of maize to chemicals and biofertilizers. *Crop Research* (Hisar). 9(2): pp. 233-237.

Moffat, A. S.1995. "Plants Proving Their Worth in Toxic Metal Cleanup," *Science*, 269: pp. 302-303.

Mo, S., Chol, D. S. and Robinson, J. W. 1989. Uptake of Mercury from aqueous solution by Duckweed: the effect of pH, Copper and humic acid. Environ. Health. 24: pp.135-146.

Mohanty, M. and Patra, H. K. 2011. Attenuation of Chromium toxicity in mine waste water using water hyacinth. *J. Stress Physiol. Biochem.* **7**: pp. 335-346.

Mukhopadhyay, S. and Maiti, S. K. 2010. Phytoremediation of Metal mine waste. *Appl. Eco. Environ. Res.* 8: pp. 207-222.

Moyukh, G. and Singh, S. P. 2005. Phytoremediation of RDX & HMX by *Ophiophogon blachei* and *Lycopersicon* sps. *Biotechnological Applications in Environmental Management*. Edited by R. K. Trivedy & Sadhana Sharma. BS publications. pp. 92-100.

Nandkumar, P. B. A., Dushenkov, H., Motto, I. and Ruskin. 1995. Phytoextraction: Use of plants to remove heavy metals from soils. *Environ. Sci. Technol.* 29 (5): pp. 1232-1235.

National Research Council (NRC), "Lead in the Human Environment," National Academy of Sciences, Washington DC, 1980, p.525.

Nazli, M. F. and Hashim, N. R. 2011. "Heavy Metal Concentrations in an important Mangrove Species, *Sonneratia caseolaris*, in Penisular Malaysia." www.tshe.org/ ea/pdf/vol3s%20p50-55.pdf.

Neelima, P. and Jaganmohan Reddy, K. 2006. Bioabsorption of some Heavy Metals in different plant species. *Nature & Environment and pollution Technology ©Technoscience publications*. 5(1): pp. 53-55.

Nirmal, I. J., Sajish, P. R. Nirmal, R., Basil, G. and Shailendra, V. 2011. "An Assessment of the Accumulation Potential of Pb, Zn and Cd by *Avicennia marina* (Forsk.) Vierh. In Vimleshwar Mangroves, Gujarat, India." http://notulaebiologicae. ro/nsb/article/viewfile/5593/5343

Nwoko, C. O., Okeke, P. N., Agwu, O. O. and Alkpan, I. E. 2007. Performance of *Phaseolus vulgaris* L. in a soil contaminated with spent-engine oil. *Afr. J. Biotechnol.* Vol. 6 (16): pp. 1922-1925.

Ojala, J. C., Jarell, W. M., Menge, J. A. and Johnson, E. L. V. 1983. Influence of mycorrhizal fungi on the nutrition and yield of onion in saline soil. *Agron. J.* 75: pp. 255-259.

Olson, P. E., Reardon, K. F. and Pilon-Smits E. A. H. 2003. Ecology of rhizosphere bioremediation. In: *Phytoremediation: transformation and control of contaminants,* ed. McCutcheon SC, Schnoor JL, Wiley NY. pp. 317-354.

Pahalawattaarchchi, C. S., Purushottaman, C. S. and Venilla, A. 2011."Metal Phytoremediation Potential of *Rhizophora mucronata* (Lam.)." http://nopr. niscair.res.in/bitstream/123456789/.../IJMS%2038(2) %20178-183.pdf

Parvaresh, H., Abedi, Z., Farshchi, P., Karami, M., Khorasani, N. and Karbassi, A. 2011. "Bioavailability and Conentration of Heavy Metals in the Sediments and Leaves of Grey Mangrove, Avicennia marina (Forsk.) Vierh, in Sirik Azini Creek, Iran," http://www.mendely.com/research/bioavailability-conentration-heavy-metals- sediments-leaves-grey-mangrove-avicennia-marina-forsk-vierh-sirik-azini- creek-iran/

Patil, G. B and Lakshman, H. C. 2003. Effect of AMF and saline water with and without additional phosphate on *Elusine coracana* (Finger millet). *Indian J. Env. And Ecopl.* 7(3): pp. 477-482.

Pattanik, R., Sahu, S., Padhi, G.S. and Mishra, A.K. 1995. Effect of inoculation of vesicular arbuscular mycorrhiza on horsegram (*Macrotyloms uniflorum*) grown in soil in iron-mine area. *Indian Journal of Agricultural Sciences.* 65(3): 186-190.

Paz-Alberto, A. M., Sigua, G. C. 2013. Phytoremediation: A Green Technology to remove Environmental pollutants. *American Journal of climate Change.*2: pp. 71-86.

Peterson, R. I. And Farquhar, M. L. 1994. Mycorrhizas – integrated development between roots and fungi Mycologia. 86: pp. 311-326.

Pierzynski, G. M., Schnoor, J. L., Banks, M. K., Tracy, J. C., Licht and Erickson, L. E. 1994. Vegetative remediation at superfund sites. Mining and its environ. Impact. Royal soc. Chem. Issues in *Environ. Sci. Technol.*1: pp.49-69.

Prasad, M. N. V., Freitas, H. M. and De, O. 2003. Metal hyperaccumulation in plant- biodiversity prospecting for Metal technology. *Electron. J. Biotechnol.* 6: pp. 110-146.

Prasad, G.K., Singh, S.K., Das, P.K. and Natha, S. (1997). Influence of VAM, macro and micromutients on vegetative propagation of *Dendrocalamus strictus. Indian Forester.* 123(9): pp. 863-866.

Preeti, P. P., Tripathi, A. K. and Shikha, G. 2011. Phytoremediation of Arsenic using *Cassia fistula* Linn. Seedling. *Int. J. Res. Chem.* 1: pp. 24-28.

Primavera, J. H., Sadaba, M. J. H. L., Labata and Altamirano, J. P. 2004. " Handbook of Mangrooves in Philippines-Panay," SEAFDEC Aquaculture Department, Iloilo.

Puspha. K. Kavatagi and Lakshman, H. C. 2011. Impact of Arbuscular mycorrhizal fungi, pressmud and Indole acetic acid on two varieties of *Lycopersicon esculentum* L. *Intern. J. Environ. Sci.* 1(1): pp. 117-121.

Qui, Y. W., Yu, K. F., Zhang, G. and Wang, W. X. 2012. Accumulation and Partitioning of Seven Trace Metals in Mangroves and Sediment Cores from Three Estuarine Wetlands of Hanian Island, China," http://www.sklog.labs.gov.cn/ article/ B11/B11012.pdf

Rajendran, P. and Gunasekaran, P. 2006. Sorption and other Remediation Strategies. *Microbial Bioremediation*. MJP Publishers. Chennai. pp. 127-128.

Rao, A. V., Tarafdar, J.C. and Tak, R. 1999a. Role of AM fungi in rehabilitation of mine spoils Abstract: *National Conference on Mycorrhiza*, Barkatulla Universtiy, Bhopal: pp. 51.

Rao, M.S. and Reddy, P.P. 2001. Control of *Meloidogyne incognita* on aubergines using *Glomus mosseae* interacted with Paecilomyces *iliacinus* and neem cake. *Namatologia Mediterranea*. 29 (2): pp. 153-157.

Rao, M.S., Srinivas, P., Reddy, G.L. and Reddy, S.R. 2002. Arbuscularmycorrhizal dependency of two forest tree species in coal mine disturbed soils. In: *Frontiers in Microbial Biotechnology and Plant Pathology* (Eds. C. Manoharachary).

Raskin, I. and Ensley, B. D. 2000. Phytoremediation of Toxic Metals: Using Plants to Clean Up the Environment. John Wiley and sons, inc., New York.

Ratageri, R.H., Taranath, T.C. and Lakshman, H.C. 2006. Toxicity of dimethoate on primary productivity of a Lentic aquatic ecosystem: A Microcosm *Approach. Bulletin of Environmental contamination and Toxicology*. 76(3): pp. 373-380.

Rauser , W. E. 1990. Phytochelatins. *Ann. Rev. Biochem*. 59: pp. 61-86.

Rauser , W. E. 1995. Phytochelatins and related peptides. *Plant Physiol*. 109: pp. 1141-1149.

Rawlings, D. E. 2002. Heavy metal mining using microbes. *Annu. Rev. Microbial*. 56: pp. 65-91.

Read, D. J., Koucheki, H. K. and Hodgson, J. 1976. Vesicular arbuscular mycorrhiza in natural vegetation systems and the occurrence of infection. *New Phytologist*. 77: pp. 641-653.

Reed, D. T., Tasker, I. R., Cunnane, J. C. and Vandergrift, G. F. 1992. Environmental restoration and separation science. In: G.F. Vandergrift, D.T. Reed, and I.R.Tasker (eds): Environmental remediation removing organic and metal ion pollutants. Amer Chem Soc. Washington DC.

Reedy, G. N. and Prasad, M. N. V. 1990. Heavy metal binding proteins, peptides: occurrence, structure, synthesis and functions: a review. *Environ. Exp*. 30: pp. 251-264.

Reeves, R. D. and Brooks, R. R. 1983. Hyperaccumulation of Lead and Zinc by two metallophytes from mining areas of central Europe. Environ. *Pollut. Ser*. A. Vol. 31: pp. 277-285

Roane, T.M. and Kellogg, S.T. 1996. Characterization of microbial communities in heavy metal contaminated soils. *Can. J. Microbiol.* 42: pp. 593-603.

Romana M Mirdhe and Lakshman, H.C. 2011. Effect of AM fungi salinity on two important legumes. *Journal of Theoretical and Experimental Biology.* 8(17 2): pp. 1-10.

Roongtanakiat, N. and Chairoj, P. (2010). "*Vetiver* Grass for Remedying Soil Contaminated with Heavy Metals,"http:/www.google.com/ Roongtanakiat+N+and+Chairoj+R=2001&meta=&aq=o&aqi=&aql=&oq=&gs_ rfai=&fp=da l b4ba 80a870679

Sakakibora, M., Aya., W. Masahiro, I., Sakae, S. and Toshikazu, K. 2007. Phytoextraction and Phytovolatilization of arsenic from As- contaminated Soils by Pteris vittata. *Proc. Ann. Int. Conf. Siols, Sediments, Water, Energy.* pp. 12-26.

Saleh, H. M. 2012. Water hyacinth for phytoremediation of radioactive waste simulate contaminated with cesium and cobalt radionuclides. Nuclear Engineering and Design. *Elsevier.* 242: pp. 425-432.

Salt, D. E., Blaylock, M., Chet, I., Dushenkov, S., Ensley, B., Nanda, P. and Ruskin, I. 1995. "Phytoremediation: A Novel Strategy for the removal of Toxic Metals from the Environment Using Plants," *Biotechnology,* 13(5): pp. 468-474.

Salt, D. E., Smith, R. D. and Raskin, I. 1998. Phytoremediation. Ann. Rev. Plant Physiol. *Plant Mol. Biol.* 49: pp. 643-668.

Santhosh, P. and Dhandapani, C. 2013. Adsorption Studies on the Removal of Chromium from Wastewater Using Activated Carbon Derived from Water Hyacinth. *Nature Environment and Pollution Technology.* 12 (4): pp. 563-568.

Saritz, R. 2005. Phytoextraction of Uranium and Thorium by native trees in a contaminated wetland. *J. Radional. Nucl. Chem.* 264: pp. 417-422.

Schildmeyer, A., Wolcott, M. Bender, D. 2009. Investigation of the Temperature-Dependendent Mechanical Behavior of a Polypropylene-Pine Composite *J. Mater. Civ. Eng.* **21** (9): pp. 460–466.

Schlegel, H.G., Cosson, J.P. and Baker, J.M. 1991. Nickel hyper accumulating plants provide a niche for nickel resistant bacteria. *Bot. Acta.* 104: pp. 18-25.

Schnoor, L. A., Licht, S. C., McCutcheon, N., Lee Wolfe., N.L.and Carreria, L. H. 1995. Phytoremediation of organic and nutrient contaminants. *Environ. Sci. Technol.* 29(7): pp. 318-323.

Schnoor, J. E. 1997. Phytoremediation Technology evaluation report TE-97-01. Ground-water remediation technologies analysis center. Pittsburgh, PA. pp. 1-16.

Schwab, A.P., Al-Assi A. A., and Banks, M. K. 1998. Adsorption of naphthalene on to plant roots. *J. Environ. Qual.* 27: pp. 220-224.

Selvapathy, P. and Sreedhar, P. 1991. Heavy metals removal by water hyacinth. *JIPHE,* India. Vol. 3: 11-17.

Shabir, H. W., Gulzar Singh, S., Haribhushan, A. Jyotsna, N., Rita, N., Brajendra

Singh, N. and Herojit, S. A. 2012. Phytoremediation: Curing Soil problems with crops. *African Journal of Agricultural Research.* 7 (28): pp. 3991-4002.

Shete, A., Gunale, V. R. and Pandit, G. G. 2011. "Bioaccumulation of Zn and Pb in *Avicennia marina* (Forsk.) Vierh. and *Sonneratia apetala* Buch. Ham. from urban Areas of Mumbai (Bombay), India," http://www.ajol.info/index.php/jasem/article/viewFile/55142/43614.

Sharma, M. P. and Adholeya, A. (2000a). Response of *Eucalyptus tereticornis* to inoculation with indigenous AM fungi in a semiarid alfisol achieved with different concentrations of available soil P. *Microbiological Research* 154(4): pp. 349-354.

Sharma, M. P., Atimanav, G., Bhatia, N. P. and Adholeya, A. 1996b. Growth responses and dependence of *Acacia nilotica* var. *cupriciformis* on the indigenous arbuscular mycorrhizal consortium of a margina wasteland soil. *Mycorrhiza* 6(5): pp. 441-446.

Singh, S., Eapen, S., Thorat, V., Kaushik, C. P., Kanwar, R. and D' Souza, S. F. 2008. Phytoremediation of [137]Caesium and [90]Strontium from solutions and low-level nuclear waste by *Vetiveria zizanoides. Ecotoxicology and Environmental Safety.* Vol. 69(2): pp. 306-311.

Sicilliano, S. D. and Germids, J. J. 1998. Bacterial inoculants of forage grasses enhance degradation of 2-chlorobenzoic acid in soil. *Environ. Toxicol. Chem.* 16: pp. 1098-1104.

Singarachary, M. A. and Girisham, S. 2002. Heavy metal tolerance of Ectomycorrhizal Fungi and Utility of metal tolerant EM Fungal strains in Reclaiming Wastelands created by Mining operations. *Bioinoculants for sustainable Agriculture and Forestry* (eds) Reddy, S.M. and Ram Reddy, S. Scientific publishers (India), Jodhapur. pp. 27-30.

Sykes, M.Y. Vina and Abubkar, S. 1999. "Biotechnology: Working with Nature to improve Forest Resources and Products," *International Vetiver Conference.* Chiang Rai, pp. 631-637.

Tangahu, B. V., Abdullah, S. R. S., Basri, H., Idris, M., Anuar, N. and Mukhlisin, M. 2011. A review on Heavy metals (As, Pb and Hg). Uptake by plants through Phytoremediation. *Int. J. Chem. Eng.* pp. 1-31.

Tauris, B., Borg, S., Gregersen, P. L. and Holm, P. B. 2009. A roadmap for Zinc trafficking in the developing barley grain based on laser capture microdissection and gene expression profiling. *J. Exp. Bot.* 60: pp. 1333-1347.

Thatoi, H. N., Padhi, G. S. Misra, A. K. 1999. Selection of efficient VAM fungi for improvement of growth and nodulation of *Acacia nilotica* growth in chromite mine waste soil. Abstract: *National Conference on Mycorrhiza*, Barkatulla University, Bhopal: pp. 54.

Thatoi, H. N., Sahu, S., Misra, A. K. and Padhi, G. S. 1993. Comparative effect of VAM inoculation on growth, nodulation and *Rhizobium* population of sababul (*Leucaena leucocephala* (Lam.) de Wit.) grown in iron mine waste soil. *Indian Forester.* 119(6): pp. 481-489.

Truong, P. N. and Baker, D. 1996. "*Vetiver* Grass System for Environmental Protection. Royal Development Projects Protectoin," Technical Bulletin No. 1998/1, Pacific Rim Vetiver, Office of the Royal Development Projects Board, Bangkok.

Truong, P. N. 1995. "Stiffgrass Barrier with *Vetiver* Grass. A New Approach to Erosion and Sediment Control," *2nd International Vetiver Conference*, 18-22 January 1995, Pechaburi.

Uera, R. B., Paz-Alberto, A. M. and Sigua, G. C. 2007. "Phytoremediation Potentials of Selected Tropical Plants for Ethidium Bromide," *Environmental Science and Pollution Research*, 14(7), pp.505-509.

United States Protection Agency (USPA) 2000. Introduction to Phytoremediation. EPA 600/R-99/107. U.S. Environmental protection Agency, office of Research & Development. Cincinnati, OH.

Velásquez L, and Dussan, J. 2009. Biosorption and bioaccumulation of heavy metals on dead and living biomass of *Bacillus sphaericus*". *J. Hazard. Mater.* 167 (1-3): pp.713–716.

Vidali, M. 2001. "Bioremediation. An overview," Pure and Applied Chemistry, 73(7): pp. 1163–1172.

Varun, M., D' Souza, R., Pratas, J. and Paul, M. S. 2012. Metal contamination of soils and plants associated with the glass industry in North Central India : Prospects of Phytoremediation. *Environ. Sci. Pollut. Res.* 19: pp. 269-281.

Vijaya, T., Srivasuki, K. P. Sastry, P. S. 1996. Role of gibberellic acid in teak seed germination and the effect of *Glomus macrocarpus* on growth and sodic soi tolerance. *Annals of Forestry.* 4(2): pp. 211-212.

Volesky Bohumil, 1990. Biosorption of Heavy Metals. Florida: CRC Press. ISBN 0849349176.

Volkering, F., Breure, A. M. and Rulkens, W. H. 1998. Microbiological aspects of sulfactant uses for biological soil remediation. *Biodegradation* 8: pp. 401-417.

Wenzel, W.W., Bunkowski, M., Puschenrieter, M. and Horak, O. 2003. Rhizosphere characteristics of indigenously growing nickel hyperaccumulator and excluder plants on serpentine soil. *Environ. Poll.* 123: pp. 131-138.

Wislocka, M., Krawczyk, J., Klink, A. and Morrison, 2006. "Bioaccumulation of Heavy Metals by Selected Plant Species from Uranium Mining Dumps in the Sudety Mountains, Poland," *Polish Journal of Environmental Studies*, 15(5): pp. 811-818.

Woodward, D., 1996. Passive gradient control. In W. W. Kovalick and R. Olexsey (eds), workshop on Phytoremediation of organic wastes, December 17-19, 1996, Ft. Worth, TX, A USEPA.

Xia, H. and Ma, X. 2005. Phytoremediation of ethion by water hycianth (*Echhornia crasspes*) from water. *Bioresource Technology.* College of Food Science, Biotechnology and Environmental Engineering, Zhiejang Ghongshang University, Hangzou.

Zadeh, B. M., Savsghebi-Firozabadi, G. R., Alikhani, H. A. and Hosseini, H. M. 2008. Effects of Sunflower and Amaranthus Culture and Application of inoculants on Phytoremediation of the Soils contaminated with cadmium. *Amer. Euras. J. Agric. Environ. Sci.* 4: pp. 93-103.

Zheng, W., Chen, X. and Lin, P. 2012. "Accumulation and Biological Cycling of Heavy Metal Elements in *Rhizophora stylosa* Mangroves in Yingluo Bay, China."

Zhang, J. E., Liu, Y. Ouyang, Y., Lia, B. W. Zhao, B. L. 2011. "Physiological Responses of *Sonneratia apetala* Buch-Ham Plant to Wastewater Nutrients and Heavy Metals." http://www.tandfonline.com/doi/abs/10.1080/1522651100367139 5#preview-physio

Zhu, Y. L., Zayed, A. M., Qian, J. H., Desouza, M. and Terry, N. 1999. Phytoaccumulation of trace elements by wetland plants: 11. Water hyacinth. *J. Environ. Qual.* 28: pp. 339-344.

INDEX